SPACES OF GEOGRAPHICAL THOUGHT

SPACES OF GEOGRAPHICAL THOUGHT

Deconstructing Human Geography's Binaries

EDITED BY PAUL CLOKE & RON JOHNSTON

London • Thousand Oaks • New Delhi

First published 2005

SAGE Publications Ltd
1 Oliver's Yard
55 City Road
London EC1Y 1SP

SAGE Publications Inc.
2455 Teller Road
Thousand Oaks, California 91320

SAGE Publications India Pvt Ltd
B-42, Panchsheel Enclave
Post Box 4109
New Delhi 110 017

British Library Cataloguing in Publication data

A catalogue record for this book is available from the British Library

ISBN 0 7619 4731 0
ISBN 0 7619 4732 9 (pbk)

Library of Congress Control Number 2004095949

Typeset by C&M Digitals (P) Ltd., Chennai, India
Printed in India at Gopsons Paper Ltd, Noida

Contents

Notes on Contributors

John Agnew is a Professor of Geography at UCLA, Los Angeles, USA. He is the author of numerous books, including *Place and Politics in Modern Italy* (2002), *Geopolitics* (2nd edition, 2003) and *Making Political Geography* (2002).

Trevor Barnes is Professor of Geography, University of British Columbia, Canada, where he has been since 1983. He is the author or editor of eight books and numerous essays. His most recent work is about geography's quantitative revolution.

Paul Cloke is Professor of Geography at the School of Geographical Sciences, University of Bristol, UK. He is Founder Editor of the *Journal of Rural Studies* and has written on geographies of nature, society and ethics. As well as recent research-based books on rural homelessness (with Paul Milbourne and Rebekah Widdowfield) and Tree Cultures (with Owain Jones), he has co-authored a series of books on philosophy, method and substantive issues in human geography (*Approaching Human Geography (1991), Introducing Human Geography (1999), Envisioning Human Geography (2004) and Practising Human Geography (2004)*) with ex-colleagues from University of Wales, Lampeter. He is currently working on projects dealing with uneven geographies of homelessness and ethical consumption.

Kevin R. Cox is Professor of Geography at the Ohio State University, USA. His interests are in the politics of local and regional development and historical geographical materialism. He is the author of four books and his work has appeared in a range of social science journals.

Mike Crang is a Reader in geography at the University of Durham, UK. He is the co-editor of the journals *Time & Society* and *Tourist Studies*. He recently edited the following books: *Tourism: Between Place and Performance* (2002, with Simon Coleman); *Thinking Space* (2000, with

Nigel Thrift); and *Virtual Geographies: Bodies, Spaces and Relations* (1999, with Phil Crang and Jon May). He has published papers on time, memory and vision in a variety of journals and is currently working on a project about the transformation of everyday space–time through electronic communications.

Nicky Gregson is Professor in Geography at Sheffield University, UK. Her research interests are in consumption and material culture and she is co-author of *Second-Hand Cultures* (2003) and *Servicing the Middle Classes* (1994). She was part of the Feminist Geographies (AWL) writing collective and is currently working on an ethnography of riding.

Lynda Johnston is a senior lecturer in the Department of Geography, University of Waikato, Aotearoa/New Zealand. Her research focuses on social/cultural and feminist geography, critical social theory and tourism. Her main interest is the formation of embodied subjectivities and the ways in which these are performed in different spaces (gay parades, spaces of sport, leisure places, home spaces). She is a co-author of *Subjectivities, Knowledges and Feminist Geographies: The Subjects and Ethics of Social Research* (2002).

Ron Johnston is a Professor in the School of Geographical Sciences at the University of Bristol. He has written widely on aspects of the history and nature of geography, including *Geography and Geographers: Anglo-American Human Geography since 1945* (6th edition, with James Sidaway, published in 2004) and the four editions of *The Dictionary of Human Geography*.

Joe Painter is Professor in Geography at the University of Durham, UK. His research interests lie in political geography, particularly the geographies of the state and governance, citizenship and democracy, and cities and regions. He is the author of numerous chapters and articles on these themes and the co-author (with Paul Cloke et al.) of *Practising Human Geography* (2004). He is currently working on a new book on the everyday political geographies of the state.

Susan J Smith was appointed to the Ogilvie Chair of Geography at the University of Edinburgh in 1990, and moved to a Chair in Geography at the University of Durham in 2004. Her work is concerned with all kinds of injustice and inequality. She has contributed to debates on citizenship and social policy, the problems of racism and gender inequality, the indignity of victimization and fear of crime, and the intractable link between housing and health. Currently, she is interested in the diverse moral economies caught between the neo-liberal performance of markets and an emergent ethics of care.

Michael Watts is currently the Director of the Institute of International Studies (IIS), and Class of 1963 Professor of Geography and Development Studies at the University of California, Berkeley, where he has taught for over twenty-five years. His current research and writing addresses the natural and social history of oil in Nigeria entitled *Black Gold: Oil, Nation and Violence in Nigeria*. He has received a number of awards and fellowships from scholarly organizations, including the Social Science Research Council, the National Science Foundation and the Guggenheim Foundation.

1 Deconstructing Human Geography's Binaries

Paul Cloke and Ron Johnston

Categorizing identity

To survive in the world we simplify it. This strategy is adopted by both individuals and collectives – although the distinction between the two is blurred because one cannot exist without the other, and so even the first sentence of this introductory chapter involves binary thinking! Our simplifications do not have to be binary, of course: binary divisions are a particular case of the wider activity of classification and categorization. But using too many categories denies the goal of simplification: psychologists argue that the average person can control her or his thinking along seven dimensions – hence the classic paper on 'the magic number seven, plus or minus two' (Miller 1956) – but most people, most of the time, use a coarser-grained structure to aid their thinking and decision-making.

Categorization – whether binary or not – is more than an aid to coping with complexity, however; it is also a means for creating our identities, again both individual and collective. Part of the determination of our essence – of our being in the world – involves defining ourselves: it is often assumed that 'we are what we are because of who we are', which involves us identifying with some type or category and, as a consequence, denying that we belong to other categories. This process of self-identification is continuous throughout our lives although it will be more intense at some times than others – when we are coming to terms with new contexts, for example, such as moving away from home for the first time, perhaps to university. As we encounter new stimuli and situations we have to place ourselves relative to them – to decide who we are; whom we identify with.

These processes of self-identification involve us joining groups and/or categories – not in the sense of applying to be a member and paying an entrance fee (although this may be the case). Many of the categories are vague and undetermined: we may identify with 'the working class' for example, a category that only exists as an abstraction – although social scientists have spent decades trying to define it precisely (and failing, because in the end it is just that – a vague abstraction – though one with a great meaning for many people).

Others are presented to us by people who encourage us to join their categories, however: political parties, for example, claim to represent particular class fractions within society, and invite those who identify with a particular class to vote for them – if not to join them and work for their goals.

Almost all of the categories that we use are 'social constructions' – they are created, not given. A few – such as the distinction between human and non-human, or between animate and inanimate, or between land and water – may be considered natural in the sense of being pre-given, or existing outside any imposed categorization involving human thought, although even these pre-given distinctions have become the subject of considerable debate. The great majority of categories are not, however, pre-given; we live in worlds of human-created categories, which we modify when new situations arise that call for new responses.

Categorizations – into binaries or more complex structures – are not just 'normal' within society, therefore: they are necessary for it. Without such simplifications, societies could not exist: they could not operate without placing people and things into categories. (Think of who qualifies for free school meals, or a subsidized university education!) As such, the study of categories and of binary thinking is central to any intellectual activity, across all disciplines: we need to know about the categories being deployed in order to appreciate the society we are studying, and we need to deploy our own categories in order to undertake that study.

Geographical binaries

For geographers, this process of categorization (binary or otherwise) is of particular interest because basic themes within our particular subject matter – environment, space and place – are frequently used in creating collective and personal identities. Many of us identify with places (some very intensely), at a variety of scales – from our home, through our street and neighbourhood, town and region, up to a nation-state. In part, we are who we are because of what places we choose to associate with (which in most cases are places we have lived, worked or played in, even if we no longer do so). And we may express that identity in a variety of ways – as with supporting sports teams linked to those places, or political parties which represent their residents. Binary positions are then often taken up – us versus them, or 'we here' against 'them there'.

Although much of this place-identification is informal, one aspect of it is not only very formal but is also to a large extent imposed upon us – and is inherently geographical. The great majority of people in the world are citizens of one of its nation-states. (Some are citizens of more than one, but they form a very small minority of the total.) To a large extent this is an aspect of our identity over which we have little control: most of us are citizens of a state because either we were born within its territory or, wherever we were born, our parents are citizens of that state and associate us with it as well. We can apply to move to another state and become one of its citizens – but that is a privilege that may be denied to us, and not a right.

This association with a state is not only central to most people's identities, it is also geographical. One of the defining characteristics of a state in the modern world-system is that it is associated with a defined – and in almost all cases demarcated – piece of territory (possession of which may be disputed by other states). As social science theorists have argued, a system of states is necessary to the successful management of contemporary modes of production (including global capitalism): without them, collective action would be impossible and anarchy would prevail. Furthermore, in order to be effective, states must have a monopoly of certain types of power: in effect, they must be able to control their citizenry in order to operate, though it is rare for such control to be effected by a state apparatus for very long without some element of popular consent. For this reason, part of the operation of the state apparatus involves encouraging people to identify positively with their state – to feel part of it, to participate in its activities and to give it a central place in their processes of self-definition.

Membership of a state is therefore part of our identity; states are necessarily territorially defined, and processes of creating one's self involve identifying with our state's territory. Part of our very nature in contemporary society thus involves geography – in particular the ever-changing geography of territorial containers within which the management of that society is undertaken. By identifying with 'our' state, we are not only accepting the power of its state apparatus over our lives but also accepting – and indicating – that we do not identify with any other state. Geographically binary thinking, therefore, underpins a core feature of our human nature.

But that binary thinking – of 'us' versus 'them' – is rarely symmetrical: usually it involves 'us' considering we are superior to 'them', and in many cases and situations being encouraged (by the state apparatus, for example) to take such a view. Difference thus becomes more than something to be exalted: it is a potential basis for conflict – as world history has demonstrated to us so many times. Much geopolitical thinking involves promoting positive images of 'us' and negative ones of 'them': think of the theory of the 'white man's burden' which underpinned much European colonization of the rest of the world, or Reagan's portrayal of the Soviet Union as the 'evil empire', or the beliefs that underpin territorial conflicts in Palestine–Israel and many other parts of the contemporary world. Identity and difference, built on geographical foundations, thus become the bases for power relationships, foment inter-group tension and, potentially, conflict.

'Them' and 'us' is, of course, an over-simplification, and our views of the world are usually more nuanced: we recognize more than two categories and behave towards each accordingly. This sometimes involves allowing for 'special cases' – as in statements such as 'some of my best friends are *x*' which allow people to avoid being in difficult circumstances when negative images of *x* are being broadcast. But when simplicity is called for, and decisions have to be made, too often we slip into binary mode: all the world is either with us or against us!

Binary geographies

Just as geography forms the foundation of our identity, so our association with the discipline of geography adds to our processes of self-identification. One of the key areas in which categorization is necessary in dealing with the complexity of the world is in its study. Knowledge and its means of acquisition are divided up into another set of containers – academic disciplines. These are clearly not 'pre-givens': they are socially created divisions which we find necessary and useful. Some might seem almost pre-given because of their subject matter, such as mathematics – what we might term *forms of knowledge*; most are not, however – they form *fields of knowledge* which are the equivalent of tribal territories, covering areas of study that for some reason (usually lost in the past) have seemed sensible to put together into a separate discipline.

Those who identify with a discipline – either temporarily (as a student) or permanently (by becoming a professional and making one's career in the discipline) – then often engage in binary thinking. This may not involve 'them' and 'us' representations – though often it does: people have 'hierarchies' of disciplines in terms of their assumed status, and these are often used in power-broking. Certainly among the professionals, their tribal territory has to be defended (which to a considerable extent means that their jobs have to be!). They want resources for their university departments, students to choose to take their courses, scholars from other disciplines not to 'invade' their intellectual niches, and so on. In academic life, just as everywhere else, we simplify by creating categories – and then people identifying with those categories come into conflict.

That happens within disciplines, too. Geography is the umbrella term usually deployed for the discipline we (the authors) work in and which you (the readers) study. But geography itself has its subdivisions – many of them binary: physical and human, for example; or qualitative and quantitative; or economic and cultural; or ... And people in those subdivisions compete too – for status and resources within the discipline, for adherents. Part of their case for support involves portraying the value of their work which, in a zero-sum game of finite resources, means that implicitly (if not explicitly – and very often it is, as any study of changes within geography over the past 60 years will show) what 'they' do has to be portrayed as inferior to what 'we' are engaged in. And yet, as we all really know, such categorizations – whether binary or slightly more nuanced – are really over-simplifications, and if the categories were brought together the cross-fertilization would bring massive benefits.

What we have identified is very much a paradox, therefore: thinking in categories – of which binary thinking is the most extreme case – is necessary to simplify the world in order to begin understanding it. But using those categories can then not only impede that search for understanding – by acting as blinkers – it can also lead to unproductive tensions between those who adhere to one of the positions and are thus opposed to the other(s). Sadly, the working-out of this paradox is rarely either easy or conflict-free: too often once territories are demarcated then people associated with them feel the need to defend them – if not attack others. Breaking down binaries is continually necessary, therefore, and this book is intended as a contribution to that continuing process.

A Binary world?

The history of human geography has been waymarked both by binary thinking and by exhortations to bridge between the philosophical and material polarities emerging from such thinking. Resultant landscapes of understanding have thereby exhibited a curiously double-edged character. Analysis and interpretation of human geographical phenomena have tended to fall easily into categories of seemingly distinctive opposition – urban/rural, economic/cultural, spatial/social, quantitative/qualitative and the like – and the professional paraphernalia of human geography, such as journals, books, courses and research specialisms and reputations, have served to render these categories more concrete. Yet alongside these categorical processes and practices there has been a naggingly consistent desire by some geographers at least to argue against the domination of polarized categories and to sponsor explorations of middle ground between them: to deconstruct the binaries and create new hybrid syntheses containing the best elements of both.

Nowhere is this double edge more clearly demonstrated than in the overarching relationship between human and physical geography, within the context of which most human geographers ply their trade. Through the second half of the twentieth century, and now beyond, the 'human' and 'physical' sections of geography have been increasingly wedged apart by the forces of differential philosophy, methodology and mindset – to a considerable extent reflecting very different (and sensible) conceptions of the nature of science. As scientific practices and reputations grow ever more specialized technically, and as the human and physical 'sides' of geography reach out in very different interdisciplinary directions thus exacerbating a social science/arts–science divide, so the forces of binary categorization have tended to desert the potential middle ground in favour of the excitement of exploration at the binary edges.

It is worth noting here that these tendencies are by no means recent. For as long as it has been argued that geography's distinctive role is in bridging the gap between physical and human phenomena, the dangers of binary tendencies have been warned against. Thus, Wooldridge and East writing in the United Kingdom in 1951 argued:

> There is a degree of a false dichotomy here – a division of the subject into 'physical geography' and 'human geography'. Such cleavage is the very thing geography exists to bridge, and it is false to its central aim wherever, and for whatever reasons it recognises two 'sides' in the subject. (p. 28)

This position was not replicated on the other side of the Atlantic, however, where it was argued that although the physical landscape was an important constraint to and opportunity for human action, appreciation of the formation of that landscape and the nature of its contemporary physical processes – how rivers erode their banks, for example – was not necessary in order to understand how people responded to their environments in the creation of 'humanized landscapes': in other words, it is not necessary to know how a stage is built in order to act on it (see James and Jones 1954). It is interesting to note, for example, that a special issue of the *Annals* to celebrate the Association of American Geographers' 75th anniversary – published in 1979 – contained no papers on physical geography, a fact responded to by the then President of the AAG, who was a physical geographer (Marcus 1979).

Indeed, this lack of interest in how the stage was built extended to some human geographers in North America at that time. Hartshorne's (1939) classic portrayal of *The Nature of Geography* – based very largely on German writings – not only took the position on physical geography just outlined but also argued that the study of historical geography was unnecessary except insofar as it threw light on the present. In this, he was very much at odds with geographers of the 'Berkeley School', led by Carl Sauer, whose own classic *The Morphology of Landscape* (Sauer 1925) took a very different view – one shared by historical geographers in the UK. Hartshorne was at least partly converted by some of his colleagues in the 1950s, and his later *Perspective on the Nature of Geography* (Hartshorne 1959) reflected this change of stance.

This brief excursion into the history of geography as an academic discipline during the twentieth century illustrates one of our earlier contentions – most binaries are human constructions and are not naturally 'pre-given'. Indeed, it goes further, because it shows that binary constructions vary over both space and time: what was accepted as a sensible division of academic activity in one place was not replicated elsewhere – and what was considered sensible there at one date had been altered 20 years later! Our binary worlds are in constant flux.

The relative balance between the power of binary classification and the purchase of bridge-building between those binaries will often depend on the ways in which attempts to reach out from the philosophical and methodological positionings of one binary serve to colonize the other and enforce ways of knowing and understanding which are incompatible with those that characterize the other. In the case of human and physical geography, for example, attempts at integration have succeeded and failed according to the compatibilities that can be achieved across the supposed divide. For Wooldridge and East, for example, physical geography was the stage on which human geography's performance is played out:

> The serious study of the subject cannot begin without the findings of physical geography; the play cannot proceed without a stage and it is a stage, be it noted, which plays a much larger part in the action than in theatrical performances. (1951: 29)

However, this endlessly possible and often subtle interconnection of stage and players has over the years become narrowed down to a series of more specific practices and ways of understanding. The subject matter of physical geography has increasingly been expressed as a series of interacting systems in which materials and processes in, say, atmosphere, geosphere, hydrosphere and ecosphere interact directly with those in others (Wolman 1992). From this perspective, interest in the 'human' has accordingly focused on, first, how individuals and societies cope with particular expressions of these interacting systems, highlighting ideas such as risk and sustainability, but encasing them in the language of environmental modelling (see Thornes 2003) and, secondly, how they stimulate, exacerbate and modify the operation of physical processes, with consequences for how those systems operate and the resulting landscapes. Even those human geographers who are literate in quantitative methodologies and are therefore technically equipped for interaction on these terms may find themselves uncomfortable with what they might view as a potential entrapment in scientized understandings and prioritizations of human behaviour.

Some human geographers adopted systems thinking as well, in line with a view that dominated the social sciences in the post-World War II decades: they argued that not only human interactions with their physical environment but also those with other humans – especially interactions over space – could be modelled in similar ways to environmental systems, producing general statements about human behaviour from which integrating theories could be derived (see, for example, Bennett and Chorley 1978; Wilson 2000.) More recently, some human geographers have suggested revitalizing the theorization of nature–society interactions which has led to a transformed appreciation of the interconnections between and hybridities of the relational agencies of non-humans, technologies, discourses and humans. These repositionings have in turn led to calls for radically new dialogues between physical and human geographers which will

> open up the space 'between' this division of geographical labours as a richly inhabited one, teaming with the stuff of which all our futures will be made. Already the subject of radical remappings in and through the technologies of life sciences, these lively spaces are fertile ground for revisiting some unfashionable geographical concerns, such as those associated with biogeography and human ecology, strengthened by the critical insights of more recent thinking in, for example, biophilosophy and anthropology. (Whatmore 1999: 260)

This opening-up very largely refers to those aspects of physical geography that focus on the animate parts of the 'non-human world' – that is, basically biogeography (perhaps also including the geography of soils). Most physical geographers concentrate on the inanimate world, however – as with work in glaciology and hydrology – for which relevant hybridities have yet to be explored in any detail.

A clear distinction between practices on either side of the human–physical binary divide within geography reflects the nature of their subject matter and associated scientific approaches. This difference was encapsulated by Sayer (1992) in his distinction between those scientific practices that assume closed systems and those that assume open systems. In closed-system thinking, the subject matter can be isolated from all other parts of a larger system (such as the global environment) – whether artificially, as in laboratory experiments, or 'naturally', in the field. The isolated system can then be treated as a machine, which always responds to the same stimuli in the same way: it obeys laws. Even when separate systems come into contact, laws still operate – but in much more complex situations (for example, as with a large machine containing a number of subsidiary machines). Such closed-system thinking cannot be assumed for humans – either as individuals or as societies – however: unlike machines (except perhaps for some computers), humans have free will, and they learn, so that they do not necessarily respond to the same stimulus in the same way every time it is encountered – indeed, they may never encounter a stimulus twice, certainly not in the same contingent circumstances (Johnston 1989).

Thus, the epistemology of physical geography, with its emphasis on searching for universal generalizations about the operation of environmental systems, is not applicable to human geography – which is not to say that some of the methods deployed in the former are not relevant in particular situations in the latter. This is one of the major misconceptions that underpins another boundary divide within human geography – between quantitative and qualitative methods. To some, because human geography cannot seek laws therefore it cannot use quantitative methods, since their sole purpose is to do just that. Such a view is strongly refuted – as recognized by Sayer (1992) in his clear distinction between extensive and intensive research. Extensive research involves searching for patterns and relationships in large populations – such as whether there is a relationship between age and turnout at general elections, or whether people living close to nuclear power plants have a greater probability of contracting leukaemia. Identifying such patterns and relationships – usually employing a quantitative methodology as the only way of handling large amounts of information – does not imply that general laws that are applicable at other times and places are being sought; the goal is just to see whether a general relationship holds in that situation alone. Extensive research cannot establish causation – why one young person abstained at an election, for example, or why somebody living near a power plant contracted leukaemia while a neighbour did not: only intensive research – focusing on the individual cases – can do that, probably using qualitative methods. Quantification is thus a powerful means to an end, not an end in itself: without it, certain patterns and relationships in human affairs might never be identified – such as the now generally accepted link between smoking, lung cancer and premature death.

Because humans operate individually and collectively in open systems, how they act in any particular circumstance will in part reflect their situation. This is exactly the same with physical systems, with immanent processes reacting to contingent conditions (on which see Kennedy 1979): how a tropical cyclone develops depends on the nature of the surface it is crossing, for example. To some extent, therefore, human and physical geography can be seen to be addressing similar problems, despite the great differences in their subject matter – how causation operates in contingent circumstances. For this, they can share a scientific approach – realism, for which Sayer (1992) has provided a clear introduction.

Realism provides a possible basis for dialogue between physical and human geography, therefore. So do other aspects of their common interests, as mapped out by Massey (1999) in terms of a possible conversation around the issue of how to rethink notions of space–time. Massey argues that one of the things which has held physical and human geography apart for so long is the role of physics in producing an assumed model of science from which particular conceptions of time and space are inherited. Rethinking these preconceptions will be a challenging prospect even for those physical geographers willing to explore beyond the boundaries of existing approaches to environmental modelling.

In these examples of bridge-building between human and physical geography, the power of the binary classification still holds considerable sway. Despite genuine and innovative attempts to build human factors into environmental models, and to re-theorize human/non-human interconnections, the middle grounds so created will typically be viewed as 'physical' and 'human' in origin and nature, although in essence this need not be the case. The power of the binary rests both in the strong categorical assumption that 'nothing can be one thing and its opposite at the same time' (Olsson 1980: 36b), and in the professional rigidities and conservatisms which ossify binary divides once established. To some extent, of course, the state of bridge-building between human and physical geographies reflects our earlier discussion of not needing to know how a stage is constructed in order to be able to act on it. This is very clear in the large body of work on resource and environmental management, often presented over the past few decades as an area where physical and human geography can be brought together. This clearly happens: human geographers interested in such matters are concerned with the physical environment, as are physical geographers with human action. But in neither case (see Johnston 1983) has there been serious and balanced interpenetration of concerns: human geographers have not felt the need to understand the physics of beach erosion when grappling with the need for coastal protection; physical geographers have not felt the need to understand why urbanization occurs to analyse the impact of high-density population centres on run-off rates. The binary remains!

Neither is the power of binary thinking bounded between human geography and 'other' subjects. Indeed the core message of this book is that human geography's take on the world is replete with ossified binaries that require serious attention. Witness, for example, our use of the binaries 'urban' and 'rural' – and the various attempts to deconstruct them (as in Pahl 1975). Conventional administrative, imaginative and intellectual boundaries have been developed between these supposedly different kinds of space, and subsequent understandings of changing spatial differentiation have served further to entrench the binary (which is sometimes resurrected in the 'popular conscience', as with claims in the UK that attempts to outlaw hunting with dogs are a consequence of urban-dwellers 'not understanding the rural way of life'). Moreover, urban/rural differences have carried with them other more hidden messages. Returning for a moment to Woolridge and East's (1951) lament for the false dichotomy between human and physical geography, it becomes clear that they view the essential geographical skill – an 'eye for the country' – as itself dichotomized between the city and the country.

> Such facility is lamentably lost by the modern tube-travelling suburban dweller; few Londoners are aware, save in the most limited sense, of the pattern of their great city, still less of the qualities and characteristics of the ground upon which it stands and of the texture and meaning of its framework of countryside. But the countryman, to this day, builds up a peculiarly full and satisfying knowledge of his physical environment ...

In these kinds of ways, the difference between urban and rural goes beyond the material look of the land and implies more deep-seated differences respectively born of a separation from and an attachment to the nature of the physical environment.

Partly as a response to such perceptions, but partly also due to the growth of a 'rural' sub-discipline which began to map out the neglect by human geographers of 'rural areas' (Cloke 1985; Clout 1972), the rural/urban divide has been kept alive by a binary model of thinking, peddling ideas of separation, difference and even opposition between the urban and its rural other. In practice, however, the divide has become blurred in all kinds of ways. Trends of counter-urbanization have brought supposedly 'urban' people into 'rural' domains, transforming the characteristics of rurality that had attracted them in the first place. In any case, traditional rural life had already been transformed by the near-universal availability of urban-based media. As Mormont (1990) has outlined, any rural/urban dualism has been completely taken over by events, and rurality has become a multiplicity of social spaces that overlap the same geographical area. Equally, the nature of recent land development has brought suburbs, theme parks, shopping centres, executive estates, industrial parks, tourist developments and the like into what were previously considered to be rural territories thus further reproducing misleading ideas about the city and the country (Wilson 1990). The rural/urban binary, however, has survived this onslaught of material reality and philosophical re-positioning. Calls to recognize the city's countryside as an integrated system (Bryant et al. 1982), to emphasize the significance of events and problems that transcend the rural/urban divide (Hoggart 1990), and to identify a 'post-rural' condition (Murdoch and Pratt 1993) have failed to de-stabilize the categorization, identity and power of urban and rural classification. Attempts to establish middle ground from the rural polarity have had little effect. Meanwhile, urban geographers have seemed not to notice any possibility of de-centring the city as a category of understanding.

Binary power?

Our starting premise in this book is that binary thinking is widespread in human geography and remains a persistent and powerful force in the shaping of landscapes of understanding (Soja and Hooper 1993). Such a dominant position is unsurprising given the place of key binaries in the history of Western thought, which has been shaped by dualistic thinking in terms of self/other, mind/body, culture/nature, male/female, active/passive, truth/illusion, and so on (see Pile 1994). Underlying human geography's binaries, however is a pattern of reason which is far from neutral in its logocentric pattern of positive and negative. By classifying the world into A/not-A, B/not-B, C/not-C categories, we commonly, though far from necessarily, ascribe power to one term over the other. As Gibson-Graham (2000) explains in terms of 'factory/not factory' distinctions in economic geography:

> The first term in the binary is endowed with positivity at the expense of the other: presence and value are attached to factory, while non-factory is absent and devalued. (pp. 97–8)

Binary structures therefore establish relations of opposition and exclusion rather than of sameness and interconnection between the two terms involved. Much of human geography, then, has potentially accepted uncritical accounts of power relations simply by endorsing binary thinking.

Over recent years, however, the unquestioning use of binaries has been put under the spotlight of close critical scrutiny. In part, this critical review has stemmed from a recognition that a whole series of binaries have lost their interpretative sharpness – Benko and Strohmayer (1997), for example, drawing on Vattimo (1988), suggest that new/old, present/past, left/right, progress/reaction, abstraction/figuration, modernism/realism and avant-garde/kitsch have all become blurred as part of the wider postmodern condition. More specifically, human geographers have sought to rediscover different understandings of identity. Thus Olsson (1980) discusses 'the law of the excluded middle' in which

> Everything is identical to itself and nothing is identical to anything else. Nothing is itself and not itself at the same time. (p. 62b)

As Cloke et al. (2004) suggest:

> Here he is underlining the assumed singularity of identity, the straightforward oneness and non-contradictoriness of the phenomena being identified, but in so doing he raises a second feature of conventional Western reasoning. This is one which tends to see things in terms of binary oppositions between supposedly dichotomous or mutually exclusive pure essences; not as condensations of many different phenomena, events and aspects fusing together in an overlapping fashion. This is not just to reiterate the finding that conventional reasoning prefers to hold things apart, assigning them to different and separate positions in the 'entity map' ..., it is also to indicate its tendency to fix identities by setting up opposites which in a sense define one another. (p. 242)

In this way human geographers have begun to view binary analyses of identity as systems of domination, arguing along with Haraway (1985) that binary traditions have been symptomatic of the logic and practice of domination of a whole list of those who are constituted as others in order to mirror the self – women, people of colour, nature, workers, animals and so on. In particular it has been feminist geographers such as McDowell (1983) who have pioneered the critical reappraisal of binary thinking, noting how core dualistic terms (A, B, C) tend to be masculinized and prioritized, while the subordinate terms (not A, not B, not C) tend to be feminized and dominated. Authors of the Women in Geography Study Group (1997) take this analysis to its logical conclusion by questioning the key taken-for-granted binary of women and men, suggesting that

> The notion that we all fit into either a male or female body is just that, notional. It cannot be sustained, either over time or space. Simply put there is nothing 'natural' about 'male' and 'female' bodies. There is nothing natural about everyone being forced into one sex or another. (p. 195)

Such an analysis leads many human geographers to move beyond binary thinking into new forms of interpretative understanding. Before discussing these moves, however, it is worth dwelling for a moment on attempts to defend, police and redevelop the idea of binary thinking. First, we should recognize that some scholars maintain the necessity of using categories to explain the social processes of identity, and that such categorization need not be a negative or regressive force. Jenkins (2000), for example, insists that

> Categorization is unavoidable in knowing the social world, and in all social identification. There is no necessary equation of categorization with stigmatization or oppression. Categorization can be positive and valorizing. (p. 20)

Susan Smith's discussion of identity politics concerning race and ethnicity (Chapter 6) presents us with a telling example of how a black/white binary can be a positive political force in the identification of and response to racism. Although not ruling out the benefits of deconstructing such binaries, Smith shows how affirmative action involving positive discrimination can flow from the positive use of binary categories. Secondly, the power or powerlessness of binary thinking can take a number of different forms. As Sayer (1991) has argued, some binaries clearly do suffer from the 'excluded middle' of which Olsson speaks, but others turn out to be more of a continuum than a dichotomy, and yet others suffer because the basic polarities of the binary are either too incoherent or too weak to bear the explanatory weight which is placed upon them. The above discussion of rural/urban in human geography reflects some of these arguments. Thirdly, it has been argued that it may not be necessary to shun the binary 'container' of understanding in order to follow the subtle effects of differences at work within that container. Sayer (1991) points out that in such cases

> The differences between entities ... are shown to be based upon a repression of the differences *within* entities ... The deconstruction of a binary opposition is thus not an annihilation of all values and differences; it is an attempt to follow the subtle powerful effects of differences already at work within the illusion of binary opposition. (p. 285)

Accordingly, for Sayer the task is to examine *in practice* the multiple meanings of binary terms so as to reveal their limitations and to promote clarifications. Such policing provides an evaluation of the realism of binary rhetoric (see Johnson 1980).

Working beyond binaries

For many, the philosophical and political flaws of binary thinking, which privileges one term over another and often represents an expression of power when the supposed opposites are in fact mutually constitutive, has led to a search for ways in which to 'escape the tyranny of dualisms' (Pile 1994: 264). One way of breaking free from binary understandings has been to pursue alternative systems of logic, developing multifaceted and fuzzy logics with which to transcend binary assumptions of either/or (Olsson 1980). Essentially, Olsson's project was to find ways of understanding certainty and uncertainty and clarity and ambiguity simultaneously. The concept of the 'fuzzy set' (that is a category with imprecise boundaries) has now become quite common in quantitative human geographies (see Openshaw and Openshaw 1997), and the wider project of entertaining certainty/ambiguity has formed the basis of philosophical innovation in human geography over recent years. Olsson's own turn was to surrealism, in order to allow crazy notions to inform the imposition of order (see Sparke 1994). Here, however, we want to give brief mention to four other moves which in different ways have sought to undo binary oppositions and take hold of certainty and ambiguity in new forms.

First, it is important to take note of the continuity in human geography of an emphasis on dialectics to resolve binary oppositions. In dialectics, according to Olsson (1980), 'every category includes both itself and its opposite' (p. 36b) and this trait has been carried through in analyses of the processes, flows and relations which give rise to systems and structures (Harvey 1996). These processes and flows operate in contradictory fashion and contain possibilities for change. They are thus capable of transforming the systems and structures they create. A focus on dialectics therefore maintains a metaphysical framework of binary opposition, but emphasizes difference as contradiction within that framework.

In similar vein, human geographers have taken to heart Giddens' (1981, 1985, 1989) project of structuration which focuses on the *dualities* of structure and agency, and argues for a non-functionalist bringing together of this supposed binary. Giddens' call is to recognize both the 'duality of structure' – that is, the manner in which structures enable behaviour, but behaviour can potentially influence and reconstitute structure – and the 'duality of structure and agency' – that is transcending the dualism of deterministic views of structure and voluntaristic views of agency. Essentially Giddens was seeking to replace the traditional dualism at the heart of social theory with a duality by claiming that

> Structure was instead implicated in every moment of action – that it was at once constraining and enabling – and that, conversely, structure was an 'absent' order of differences, present only in the constituting moments of

> interaction through which it was itself reproduced or transformed. (Gregory 1994: 112)

Thus the subjugation of the secondary term of a binary by the primary may be reconceived in terms of each term being able, simultaneously, to enable and constrain the other. As with dialectics, this idea of duality maintains the metaphysics of binary relationship as a two-way street of shaping and being shaped (see Nicky Gregson's chapter on structure and agency).

Others wishing to detach further from a framework of binaries have become interested in opening up a 'third' ground between seemingly polar binaries. This project can be pursued via the use of 'third terms', for example by focusing on sexuality to recognize alternative possibilities and experiences from those suggested by a simple binary analysis of man/woman. In some cases, third terms can simply point to the middle ground of a continuum between polarities – the longstanding categorization of 'suburban' and 'environmental', for example, serve this purpose in urban/rural and physical/human distinctions respectively. However, the use of third terms to inspire non-binary analysis potentially goes far beyond a mere elucidation of middle ground. As Lynda Johnston (Chapter 7) demonstrates, studies of sexuality have highlighted the importance not only of heterosexual and homosexual spaces, but also of the notion of 'camp', which interconnects such categories in new ways.

Use of third terms links with a significant turn towards the critical potential of 'third spaces' (Bhabha 1990; Soja 1996). Third space refers to spaces that transcend what is produced by binary processes. So whereas dividing the world for example between the rich and exploitative West and the poor and exploited remainder, has produced interesting and politically important geographies of development, it has been possible to challenge this binary distinction in such a way as to recognize more complex post-colonial spaces in which new forms of otherness can emerge and different kinds of identity can be acted out. Third spaces thus combine the material and the symbolic to elude the politics of polarized binaries and to enable the emergence of radical new allegiances by which old structures of authority can be challenged by new ways of thinking and new emancipatory practices. Pile (1994) argues strongly for the usefulness of third space. It permits: a positioning of socially constructed binaries within contexts of dissimilarity; an appreciation of how the values established by power-laden binaries can be decoded and challenged; a recognition of the potential which exists beyond the interconnecting power of binaries; and an understanding of the fluidity of space when thinking through 'the ways in which the fabric of this third space is continually fragmented, fractured, incomplete, uncertain and the site of struggles for meaning and representation' (p. 273).

These fluid characteristics of third space reflect a dissatisfaction with the fixities produced by binary thinking which is also at the heart of the move towards *deconstruction* in social science. The adoption of Derrida's (1967) post-structuralist approach has sought to confirm the creation of meaning as an unfinished process and as a space in which alternatives can be generated through political struggle. The practice of deconstruction works directly against binary metaphysics of presence and power. Binaries operate by encoding a form of master signifier which in turn exerts a certain kind of stability on the politics of difference. Deconstruction unpicks this overarching structure of signification and presents a philosophical means by which human geographers can reconsider the notion of difference beyond the fixity of binary and hierarchical restrictions, and engage in the process of developing radically heterogeneous spaces. Merely to reverse the value and power of binaries, for example by espousing the rights of woman over man, or black over white, can leave the fixed structure of understanding intact. As Gibson-Graham (2000) argues, it is more potent to

> blur the boundaries between the terms highlighting similarities on both sides of the divide, undermining the solidity and fixity of identity/presence, showing how the excluded 'other' is so embedded within the primary identity that its distinctiveness is ultimately unsustainable. (p. 99)

Deconstruction, then, recognizes that incorporation of excluded others within binary frameworks serves to familiarize a particular form of 'other' identity and thereby hollow out the meaningfulness of that identity (Doel 1999a,b). Instead, this approach points us to moments and sites of contradiction, indecision and deferral of meaning, and leads us to argue that no overarching sign can adequately convey structures of meaning which will be continually compromised by ever-changing and incomplete texts and contexts.

The move to deconstruction, then, replaces binary thinking with more fluid conversations within which new languages and politics of identity and possibility can emerge. For some human geographers, these conversations are most significantly focused on the performativities of being and becoming. Thrift (2000), for example, drawing on Deleuzian ideas of smooth and striated space, points us to the multiplicity of space–times, urging attention to how events are shaped as they happen and seeking to distance us from 'the vapid certainties of so much current cultural work' by seeking to understand 'what is carried in and carried away by different voyagers and beings in becoming' (p. 215). His emphasis is on effectivity rather than representation, on the how rather than the what, and he regards the flexibility and instability of space as a virtue of potential, reflecting endless possibility in a fleeting, dialogical and risky context which in its purest form cannot be written about, or stabilized.

It will be evident that these four moves beyond binary frameworks – dialectics, dualities, third spaces and deconstructions – reflect strongly differentiated degrees of willingness to dispense with the stuff of which binaries have been made. There is also significant variation in the ways in which political-economic materialities are retained/released, focus is dominated/shared by the cultural, and emphasis is placed on the representative/non-representative. Clearly a wish to move beyond binary thinking in human geography has itself broken free from any overarching political and ethical projects, and from any sense of certainty about how to research or even apply our understandings in the material world out there. There is perhaps a danger here that new binaries are being introduced surreptitiously, for example between the cultural and the political/economic, the representative and the non-representative, the historical and the momentary, the writable and the unwritable, the ethical and the post-ethical, the knowable and the unknowable. If so, the very fluidity of deconstruction in human geography will only have served to fuel moments of reconstruction, albeit using different frames of reference and understanding.

About the book

In this book we have asked authors to write about a particular binary associated with the construction of human geographic thought. We are interested both in the ways in which human geographers have made specific use of broader conceptual materials in order to produce 'geographical' concepts and ideas, and in the ways in which such usage has often taken a particular shape – that of working with seemingly bilateral or binary ideas in order to map out the philosophical and conceptual territories of the subject. The book therefore focuses specifically on the bilateral dialogues which have characterized the formation of human geographical thought. Such binaries are clearly not necessarily dichotomous. What marks them out as significant is the way in which they have become co-constitutive, the one usually being discussed by direct reference to the other.

Our aim, therefore, has been to bring together a collection of carefully constructed essays on the key conceptual binaries which have been critical in the ways in which human geography has been, and continues to be theorized. The essays both chart the significance of a particular binary in the co-constitution of geographic thought, and demonstrate how the polar significance of binary elements has been deconstructed in more recent attempts to unsettle the meta-narratives of modernist philosophies in human geography. Individual authors have been given considerable latitude in their task, and the resultant chapters will therefore reflect both that authorial individuality and a sense in which some binaries have been more prone to particular forms of deconstruction than others.

We do not claim that the list of chapters represents a complete suite of binaries in human geography. Indeed, one chapter that was due to be included was not delivered, leaving at least one omission. Equally, the book has taken some time to produce. Some authors produced their chapters somewhat earlier than others, and have had to wait some considerable time for other chapters to come in. We apologize to you for this frustrating but unavoidable delay. However, we believe that the book marks out a significantly different cut across most existing narratives of philosophical, theoretical and conceptual thinking in human geography, and we trust that it will contribute in some small way to a continued liveliness of debate and dialogue in our subject.

References

Benko, G. and Strohmayer, U. (eds) (1997) *Space and Social Theory: Interpreting Modernity and Postmodernity*. Oxford, Blackwell.

Bennett, R.J. and Chorley, R.J. (1978) *Environmental Systems: Philosophy, Analysis and Control*. London, Methuen.

Bhabha, H. (1990) 'Interview with Homi Bhabha: the third space', in J. Rutherford (ed.), *Identity, Community, Culture, Difference*. London, Lawrence and Wishart. pp. 207–21.

Bryant, C., Russwurm, L. and McLellan, A. (1982) *The City's Countryside*. London, Longman.

Cloke, P. (1985) Whither rural studies?, *Journal of Rural Studies* 1: 1–9.

Cloke, P., Cook, I., Crang, P., Goodwin, M., Painter, J. and Philo, C. (2004) *Practising Human Geography*. London, Sage.

Clout, H. (1972) *Rural Geography: An Introductory Survey*. Oxford, Pergamon.

Derrida, J. (1967) *Of Grammatology* (trans. G. Spivak). Baltimore, MD, Johns Hopkins University Press.

Doel, M. (1999a) *Poststructuralist Geographies: The Diabolical Art of Social Science*. Edinburgh, Edinburgh University Press.

Doel, M. (1999b) Deconstruction on the move: from libidinal economy to liminal materialism. *Environment and Planning A* 26: 1041–59.

Gibson-Graham, J.K. (2000) 'Poststructural interventions', in E. Sheppard and T. Barnes (eds), *A Companion to Economic Geography*. Oxford, Blackwell.

Giddens, A. (1981) *A Contemporary Critique of Historical Materialism, Vol. 1: Power, Property and the State*. London, Macmillan.

Giddens, A. (1985) *A Contemporary Critique of Historical Materialism, Vol. 2: The Nation-State and Violence*. Cambridge, Polity Press.

Giddens, A. (1989) *A Contemporary Critique of Historical Materialism, Vol. 3: Between Capitalism and Socialism*. Cambridge, Polity Press.

Gregory, D. (1994) *Geographical Imaginations*. Oxford, Blackwell.

Haraway, D. (1985) 'A manifesto for cyborgs: science, technology and socialist feminism in the 1980s', in L. Nicholson (ed.), *Feminism/Postmodernism*. London, Routledge.

Hartshorne, R. (1939) *The Nature of Geography*. Lancaster, PA, Association of American Geographers.

Hartshorne, R. (1959) *Perspective on the Nature of Geography*. Chicago, Rand McNally.

Harvey, D. (1996) *Justice, Nature and the Geography of Difference*. Oxford, Blackwell.

Hoggart, K. (1990) Let's do away with rural, *Journal of Rural Studies* 6: 245–57.

James, P.E. and Jones, C.F. (eds) (1954) *American Geography: Inventory and Prospect*. Syracuse, NY, Syracuse University Press.

Jenkins, R. (2000) Categorisation: identity, social process and epistemology. *Current Sociology* 48: 7–25.

Johnson, B. (1980) *The Critical Difference: Essays in the Contemporary Rhetoric of Reading*. Baltimore, MD, Johns Hopkins University Press.

Johnston, R.J. (1983) Resource analysis, resource management and the integration of human and physical geography. *Progress in Physical Geography* 7: 127–46.

Johnston, R.J. (1989) 'Philosophy, ideology and geography', in D. Gregory and R. Walford (eds), *Horizons in Human Geography*. Macmillan, London. pp. 48–66.

Kennedy, B.A. (1979) 'A naughty world', *Transactions, Institute of British Geographers* NS4: 550–8.

McDowell, L. (1983) 'Towards an understanding of the gender division of urban space', *Environment and Planning D: Society and Space* 1: 59–72.

Marcus, M.G. (1979) 'Coming full circle: physical geography in the twentieth century', *Annals of the Association of American Geographers* 69: 521–32.

Massey, D. (1999) 'Space–time, "science" and the relationship between physical geography and human geography', *Transactions, Institute of British Geographers* 24: 261–76.

Miller, G.A. (1956) 'The magical number seven, plus or minus two: some limits on our capacity for processing information', *The Psychological Review* 63: 81–97.

Mormont, M. (1990) 'Who is rural? Or how to be rural: towards a sociology of the rural', in T. Marsden, P. Lowe and S. Whatmore (eds), *Rural Restructuring*. London, David Fulton.

Murdoch, J. and Pratt, A. (1993) 'Rural studies: modernism, post-modernism, and the "post-rural"', *Journal of Rural Studies* 9: 411–27.

Olsson, G. (1980) *Birds in Egg/Eggs in Bird*. London, Pion.

Openshaw, S. and Openshaw, C. (1997) *Artificial Intelligence in Geography*. Chichester, Wiley.

Pahl, R. (1975) *Whose City? And Other Essays*. Harmondsworth, Penguin Books.

Pile, S. (1994) 'Masculinism, the use of dualistic epistemologies and third space', *Antipode* 26: 255–77.

Sauer, C.O. (1925) *The Morphology of Landscape*. Berkeley, CA, University of California Publications in Geography, 2: 19–54.

Sayer, A. (1991) 'Behind the locality debate: deconstructing geography's dualisms', *Environment and Planning A* 23: 283–308.

Sayer, A. (1992) *Method in Social Science: A Realist Approach*, 2nd edn. London, Routledge.

Soja, E. (1996) *Thirdspace*. Oxford, Blackwell.

Soja, E. and Hooper, B. (1993) 'The spaces that difference makes: some notes on the geographical margins of the new cultural politics', in M. Keith and S. Pile (eds), *Place and the Politics of Identity*. London, Routledge.

Sparke, M. (1994) Escaping the herbarium: a critique of Gunnar Olsson's 'Chiasm of thought and action', *Environment and Planning D: Society and Space* 12: 207–20.

Thornes, J.B. (2003) 'Geographers and environmental change', in R. Johnston and M. Williams (eds), *A Century of British Geography*. Oxford, Oxford University Press.

Thrift, N. (2000) 'Afterwords', *Environment and Planning D: Society and Space* 18: 213–55.

Vattimo, G. (1988) *The End of Modernity. Nihilism and Hermeneutics in Post-Modern Culture*. Cambridge, Polity Press.

Whatmore, S. (1999) 'Geography's place in the life-science era?', *Transactions, Institute of British Geographers* NS24: 259–60.

Wilson, A. (1990) *The Culture of Nature*. Oxford, Blackwell.

Wilson, A.G. (2000) *Complex Spatial Systems: The Modelling Foundations of Urban and Regional Analysis*. Chichester, John Wiley.

Wolman, G. (1992) 'Contemporary value of geography: applied physical geography and the environmental sciences', in A. Rogers, H. Viles and A. Goudie (eds), *The Student's Companion to Geography*. Oxford, Blackwell.

Women in Geography Study Group (1997) *Feminist Geographies, Explorations in Diversity and Difference*. Harlow, Longman.

Wooldridge, S. and East, G. (1951) *The Spirit and Purpose of Geography*. London, Hutchinson.

2 Agency: Structure

Nicky Gregson

Bombs, elephants, histories and geographies

> Everything he owned was in that place, all that he had ever worked for; a lifetime's accumulation of labour stored as a single cache of wood. He thought of the elephants and the bombs falling around them; the flames leaping from the well-stacked wood; the explosions; the trumpeting.
>
> It was he who had concentrated all his holdings in this one place – that too was part of the plan – and now the bombs had claimed it all. But it didn't matter; nothing mattered so long as Neel was unharmed. The rest were just things, possessions. But Neel ...
>
> (Amitav Ghosh, *The Glass Palace*, 2000: 463)

The above constitutes one of the many critical moments in Amitav Ghosh's powerful post-colonial epic, *The Glass Palace*. Here the central figure and patriarch, Rajkumar, a self-made capitalist whose wealth is bound up with teak, has gambled all: his intention is to sell all his year's accumulation of timber to contractors working on the Burma–China road. At stake is the survival of his business. But it is 1942; wartime. The Japanese are bombing Rangoon. And corralled in Rajkumar's timber yard are numerous elephants vital for the movement of the wood. But bombs and elephants are a dangerous juxtaposition, and the explosions produce terror and panic in these highly sentient creatures. In the confined space of the yard the elephants reap devastation: not only is the wood engulfed in fire but the elephants' actions bring about the death of Rajkumar's eldest son, Neel. At a stroke then, Rajkumar loses both his son and his business, and shortly afterwards – along with millions of other Indians in Burma – he, his wife Dolly, daughter-in-law Manju and baby grandchild begin the walking exodus from Burma to Bengal.

I begin this chapter with Ghosh because it seems to me that what he narrates so effortlessly yet eloquently are precisely my concerns here. That is, running seamlessly through this moment are two parallel readings of agency/structure. The one is an explication, indeed a model one, of Marx's famous dictum that 'men [*sic*] make history but not in the conditions of their own choosing'. Agency here is human agency, and an agency that is understood as intentional, purposive, meaningful action/s; structure is the conditions that simultaneously enable, frame, suggest and constrain action/s. We can see both clearly in the portrayal of Rajkumar above. On the one hand Rajkumar has plans for his business – a classic instance of intentional action. But on the other, these plans are themselves moulded and shaped by conditions. The Burma–China road only becomes significant for strategic reasons to do with the entry of the USA to the war, and as we have seen plans are reduced to ashes by the Japanese bombing of the Burmese capital. Conditions then, specifically here the geopolitical context of India, Burma and Malaya in World War II, provide both possibilities for action/s and the limits to the possible. And through all this history and histories – colonial and post-colonial – are being unmade and made.

At the same time though Ghosh conveys a very different sense of agency/structure in this extract. Indeed, there is a strong sense that the main agents here are not people but bombs and elephants, and their juxtaposition in time and space. Exploding bombs bring about the fires that devour the wood, Rajkumar's capital, charging elephants, the death of a son. And there is another sense of structure discernible here. Provisional and constituted through networks of effectivity, this is located not in a prior set of conditions, nor in institutions or rules, but in the connections brought into being through actions' effects. The effect of the conjuncture of bombs with elephants then is traumatic. With nothing left to keep the family in Burma, bombs and elephants could be argued to have set in chain a series of actions that result in trans-continental migration, and eventually personal renewal and closure; actions that link together familial histories in geographies that connect India with Burma but in ways that fold time and space back upon one another. Although never making the point – why should he need to! – Ghosh's narrative can be read as an articulation of actor-network theory.

In the remainder of this chapter I want to hold on to the doubleness of Ghosh's writing of agency/structure. I want to argue for a human geography that can admit simultaneously these two readings of agency/structure; that can run them in parallel, entwine them as Ghosh does, rather than pose them as either/or alternatives. As I argue in the following section, the history of human geography, at least over the past 20 years or so, has been one in which particular articulations of the agency/structure binary have prevailed over others. Correspondingly, it is possible to discern movement within this same time frame: from a Marxist reading to actor-network inspired accounts for example, and from an emphasis on a pre-existing, constraining sense of structure to an increasing preoccupation with a form of agency that extends beyond human agency. Such oscillations, however, have their downside. In focusing attention on one particular articulation of agency/structure and on one half of what is a binary pairing, much of the geographical literature has both promoted an orthodoxy and reduced its analysis of a binary to the singular. Not only is this problematic in itself, it also positively encourages oscillation. In contrast, I want here to attempt to break out of this pattern. To do this seems to me to require that we think continually in terms of agency/structure and that we disrupt our tendency to articulate this binary in one way. In what follows, and particularly in the third and fourth sections, I attempt to entwine the two readings of agency/structure outlined here through two examples. The two examples are not idly chosen. Together they comprise two of the most significant occurrences of 2001 in the UK and beyond. They are: the discovery of 'foot-and-mouth' disease at a pig farm at Heddon-on-the-Wall, Northumberland, near where I live, and the moment when a US civilian aircraft became a flying bomb – September 11. What these, in their differing ways harrowing, events permit me to uncover is that whilst we can and clearly do need to think of an agency that is more than human and a structure that is provisional and emergent, this cannot and should not mean that we dispense with thinking about a form of human agency that is conditional and a sense of structure that acknowledges (and contests) both the durability and the power of certain relations and configurations.

A story, some moments and a plea: or an argument for how and why human agency continues to matter

This section begins with a story. Not a 'fairy story', nor fiction, but a linear narrative of sorts about a discipline, geography. Narratives like these are important. They constitute meaning frameworks; ways in which we make sense of, indeed order, the events and particularly writings that are such a central part of academic practice. When we write these stories then we are literally making disciplines and/or knowledge bundles; constituting and imposing a retrospective order where the practice is far, far messier. Sometimes though – and texts like this one are one such – writing a mess is less than helpful. So I begin thus, but what I centre in this story is a couple of moments: one 20 years ago, the other now.

During the early years of the 1980s the agency/structure binary occupied the centre ground of theoretical debate and discussion in human geography (see, for example, Gregory 1981; Thrift 1983). Taking its cues from the then expansive explorations in social theory being produced by Anthony Giddens (1979, 1981, 1984), and specifically from his 'structuration theory', this work – about which there is surely a sociology to be written – sought to transcend the twin poles of voluntarism and determinism; to see both agency and structure as brought into being in action, in the moment of their instantiation. Looking back at this work – admittedly a far from straightforward exercise – much of it now seems a rather tortuous elaboration of Marx's historical materialism bolted on to a rigorous theoretical analysis of individual agents and agency. But what it did inspire, and indeed introduce into the orbit of geography, was an awareness of a range of studies that exemplified many of the core tenets of this position: 'classics' like Philip Abrams' *Historical Sociology* (1983) and E.P. Thompson's *The Making of the English Working Class* (1963). Here, for example, in an exemplary passage, Abrams articulates with supreme clarity the core paradox of the agency/structure binary:

> All varieties of sociology stress the so-called 'two-sidedness' of the social world, presenting it as a world of which we are both creators and the creatures, both makers and prisoners; a world which our actions construct and a world that powerfully constrains us. The distinctive quality of the social world for the sociologist is, accordingly, its *facticity* [emphasis in original] – the way in which society is experienced by individuals as a fact-like system, external, given, coercive, even while individuals are busy making and re-making it through their own imagination, communication and action ... The two-sidedness of society, the fact that social action is both something we choose to do and something we have to do, is inseparably bound up with the further fact that whatever reality society has is an historical reality, a reality in time. When we refer to the two-sidedness of society we are referring to the ways in which, in time, actions become institutions and institutions in turn are changed by action. (1983: 2)

And, whilst both Abrams and Thompson may have disagreed over the relation of history and sociology (see, for example, Thompson 1978), there is little that differentiates the above position from Thompson's famous claim that 'the working class did not rise like the sun at an appointed time. It was present at its own making' (1979: 9).

Although it is perhaps salutary to acknowledge and reflect on the fact that it is the latter texts – and not the veritable forests devoted to theoretical exegesis – that have lasted, the point that I want to make about this work here is about its own geographical moment. At the same time as it was endeavouring to transcend a particular theoretical binary then this work had its own disruptive effects – notably on the voluntarist and determinist readings of this same binary that then prevailed and co-existed in the geographical literature (see, for example, Harvey 1982; Ley and Samuels 1978; Massey 1979, 1984; Tuan 1976). Indeed, not only did this work overtly challenge the one-sidedness of these respective readings, but in forwarding an account based on moments of instantiation it brought into question the merits of binary conceptualization itself. In short, it both destabilized particular readings and the ontological basis for these readings, and – even though never quite making the radical, transcendary break that it potentially promised (how could it be given its terms of reference?) – it at least pre-figured some contemporary theoretical preoccupations in geography: for example, action, moments and events (Thrift 2000). Human geography then was never quite the same again. A radical claim perhaps, but one that possibly does need to be made since it is more commonplace to attribute this sea-change to the exclusive effects of post-modernism and/or post-structuralism.

A moment though it was. Lost. Forgotten. Buried. Instead, the 1990s were characterized by at times acrimonious exchanges as geography, along with the humanities and social sciences, both resisted and embraced a plethora of 'new' thinkers: Foucault, Butler, Derrida, Irigaray, Deleuze and Guattari ... Yet, amidst all this tumolt, agency/structure was a binary that never quite went away. So, in the reworking of 'structure' as discourse (I want to leave aside whether this was actually the case!), it was the denial of 'agency', or the scope for an agency outwith discourse, that was raised frequently as an objection – with the effect that the binary itself came to be re-instated. Agency however still meant human agency, and this was an understanding that continued to be connected to language and to communicative intent.

The latter is of profound importance to 'what happened next?'. For, in what amounts to one of the most critical theoretical steps of recent years we have seen the increasing acknowledgement of the agency of things; that is, the forwarding of an argument that agency is not a property unique to human beings, and that correspondingly it cannot and should not be identified exclusively in terms of linguistic and communicative capabilities and possibilities. These arguments have two origins: one is actor-network theory (ANT) – a set of positions that has received considerable exposure of late in the geographical literature (Bingham 1996; Murdoch 1997, 1998; Whatmore 1997); the other is material culture studies. Some illustration is in order.

Although there are now a few examples in the literature of geographers working with ANT (see, for example, Whatmore and Thorne 2000), some of the best instances of this lie elsewhere, with work being conducted in sociology and science and technology studies. Here the work of John Law, Annemarie Mol and, of course, Bruno Latour (1987, 1993) has been pivotal. And it is some of the key features of their work that I want to highlight here because what this engages with consistently is assemblages, networks of the animate and inanimate.

Law's (2000) account of the explosion of an Olympus 22R aeroengine on a test bed at Filton, Bristol in 1962 is particularly noteworthy. What Law does here is to move right away from a conventional, linear narrative account – the type of writing with which I deliberately began this section. So he is not concerned with uncovering, documenting and excavating the search for causes and their eventual remedies, that is with an archaeological approach and its instrumental resolution. Rather, his concerns are with the agency of this explosion. And particularly with the effects of this explosion on barriers, practices and regulations. What Law succeeds in demonstrating here is the ways in which the explosion – one specific, calamitous event – broke down the barriers that framed a specific scientific project (the TSR2), as well as those between this project and Parliament and between the project and public debate. Correspondingly, so he argues, much of the 'work' that went on following the explosion was not just about establishing scientific causes and solving these, but about struggling to restore and rebuild previous barriers. In short, to re-make order by reinstating practices and regulations in a series that located the political as prior to the administrative and technical. What we see here, and in other recent ANT accounts, is still a view of agency that is as effect, distributed through an heterogeneous arrangement of materials rather than the intentional activity of human subjects (Law 1986). But what we also see is an attempt to move beyond some of the earlier criticisms of ANT that highlighted its inattention to hierarchical relations of difference and its inclusion of everything – from door closers to scallops and human beings – within a networked performance of agency (Callon 1986). In moving attention to how and when networks fail then, ANT has managed to admit into its accounts the struggle to reconstitute the spatialities of particular networks. And it has highlighted the importance of boundaries and their maintenance to these. But, as others have commented, it still needs to find ways of accommodating relations of Otherness within networks (Lee and Brown 1994; Hetherington 1997); a move which many see as attendant upon dispensing with the spatialities of the network altogether (Mol and Law 1994; Law and Mol 2001; Hetherington and Law 2000; Law and Hassard 1999).

At first sight there is much in common between the arguments above and some current and ongoing work in material culture. In a recent edited volume, for example, Daniel Miller (2001) explores a novel take on 'estate agency'. Here, and in contrast to the majority of his collaborators, Miller examines the agency of the house (and see too Birdwell-Pheasant and Lawrence-Zúñiga 1999). Rather than see the house as expressive of individuals' agency, he considers three elaborations of 'estate agency'. One is the tradition of the English haunted house, where the figure of the ghost (benign or not) works to partially anthropomorphize the history of the house and to diminish the significance of the present occupants, who represent no more than a superficial trace in its biography. A second is an engaging personal reflection on home decoration; something that highlights how living in a particular aesthetic exerts a very specific and intimidating effect on someone who sees himself as aspiring to this representation but as unable, in a sense, to measure up to the strictures of its 'taste'. Thirdly and relatedly, Miller outlines how the temporality of particular houses themselves exert agency; how people living in 'old' and 'new' houses come to be possessed by feelings of morality, of what ought to be, in terms of interior decoration. Is it appropriate to put uPVC windows and doors in an 'old' house? Should a modern house be cluttered up with antiques?

Similarly, in some of my own work with Louise Crewe (Gregson and Crewe 2002), we have been concerned with the biographies of things and their agency. Objects like second-hand clothing for example, can be 'haunted' in ways that echo the agency of the house. Clothes that have been worn bear the traces of those who have inhabited them – their smell, their shape, even their bodily leakages (perspiration, blood ...) frequently remain, trapped between the weft and the weave (Gregson et al. 2000). And much of the consumption work that occurs after acts of second-hand clothes purchasing is about trying to erase this (often unknown) ghost. Washing, ironing, placing things in freezers or outside on a frosty night, even dry cleaning, all are rituals that are not just about personalization and possession but about divestment too. So, it is not just that purchasers exercise agency in relation to acts of clothing purchase, but the clothes too exert effects; they suggest a morality of practice.

And yet, notwithstanding that arguments like these acknowledge the agency of things and even, as in Miller's chapter, make a point of emphasizing that to do so is not, as was previously thought, a category mistake, there is a difference between them and ANT accounts. These differences are there to see in the second half of Miller's chapter, where he goes on to explore various examples of the ways in which objects can stand for both people and relationships, notably the ways in which those living in council housing in the UK are frequently 'haunted' by their belonging to the state. Here the frustrations of daily life and the limits placed on change lead not only to alienation but to a transference of agency onto 'the council' – seen as an uncaring, distant presence that possesses what the occupant cannot (the house). What we see played out here is not just the interchangeability of values between objects, people and relationships but strong overtones of Miller's dialectical approach to material culture, objectification (Miller 1987). In this, drawing on Hegel, Miller defines objectification through the processes of externalization and sublation, involving respectively the distancing of goods (objects and services) from the conditions and institutions of their production, and their reappropriation by society/ies through social subjects. This recursive position is, of course, at a considerable ontological remove from the cohabitive world of ANT accounts, where no separation between object and subject worlds is sustained.

The current moment then in human geography is one where agency is a primary concern, but one where fundamental ontological differences exert critical effects. So, whilst there is a degree of agreement over how we might answer the question 'who or what might be an agent' – literally anything, animate and inanimate – and over whether things can be and act in ways that have the potential to exceed that which we intend of them – that agency is about effectivity – for some, and here I would include myself, conceptualizations of agency still have to work with the difference that human agency makes. To extend an understanding of agency to objects is not to deny that there is such a thing as human agency (cf. Thrift 2000). Rather we have to think carefully about whether the latter is no more than a particular form of effectivity, or more than this. On balance, I happen to think that the latter is the more tenable position.

My reasons here take us back to the quote with which I began, and to Ghosh's articulation of Rajkumar's thoughts. In his moment of panic all that matters, all that has meaning, for Rajkumar is the well-being of his son; the rest is 'just things'. Hardly a novel observation, for this is one that transcends societies and cultures: the social relationship, and here it is significant that this is a kinship relation, precedes the object relation. It is quite simply of greater importance. And indeed, if we examine much of the work conducted by geographers and others on consumption practices recently (see, for example, Clarke 2000; Miller et al. 1998; Miller and Slater 2000), we can find a great deal of support for this position: people use things to make social relationships (a situation in which human intentionality and meaning is given precedence over objects), as well as acknowledging the effectivity of things on their actions. So, when we look at what people actually do and how they make sense of the worlds in which they live, we find that a lot of this is an elaboration of objectification, and that – correspondingly – people continue to regard the worlds of things and human subjects as distinct and hierarchically differentiated, even if they conjoin them in practice.

I remain sufficiently committed to an empirical social science practice to think that this actually matters, a lot. If people sustain through their practices (and in the ways in which they talk about these) a separation between the subject and object worlds, and between the agency of things and their own acts of intentionality and meaning, then surely we have to continue to take this seriously – or risk being dismissed as completely 'wacky', 'off the planet' or, and potentially far worse for social scientists if not artists, as an 'irrelevance'. So perhaps this is both a plea and a warning. To use a personal example: as I write one of my cats is currently miaowing on the dustbin (clearly he is signalling his intention to be let in!). Another has spent the morning, as he often does, indicating that he too would like to go out. At one level both have agency, but unless I do something about this – which would involve putting down this pen, interrupting a train of thought, and subordinating my work demands albeit only momentarily – nothing will alter their current locations; the current arrangement of things will persist. Bottom line, I get to control their spatial explorations, however much they might attempt to make things otherwise! And to talk about this in any other way would be bordering on the ridiculous for there is a hierarchical relation at work here. Seductive though the extension of agency to objects and things may be then, to take this to its radical conclusion and to deny the way in which people continue to privilege in practice a human subject connected to human agency is to attempt to rearrange matter in ways that perhaps reveals rather more concern with the reordering of knowledge boundaries (science/social science) than with the more mundane but none the less important tasks of documenting, interpreting and accounting for the practices that constitute everyday social lives. In Geography, as in other social sciences, the accounts we construct are themselves frameworks of meaning; situated frameworks, impositions of order that make sense of worlds and realities. But to make sense these accounts need to be not just intelligible but to correspond with the worlds and realities they purport to describe, interpret and explain. If our accounts fail to connect then they risk being construed as valueless; food for thought perhaps.

I want to end this section though by turning away from moments; to recall the sense of narrative that I started with. Now, whilst the vast majority of geographical writing at any time is bound up almost exclusively with moments – particular practices, occurrences and configurations of time/space – these concerns can be, and frequently are, conjoined; drawn together through narratives. Moments are woven together as 'events', often in a linear sequence or series, to produce disciplinary histories (Cloke et al. 1991; Livingstone 1992). And, in a very small way, this is what I started to do above. At a much broader scale though – in sweeps of 'grand' historical writing of the type characteristic of the Annalistes, or of Braudel or Wallerstein (Braudel 1949, 1979; Le Roy Ladurie 1978; Wallerstein 1974) – we see a similar concern with narrative, but one that has at its core a desire to connect the historical narrative of events to human action. We see the same intent in Ghosh's novel. At the same time as the events that history has designated as critical to the course of World War II in Burma, Malaya and India march sequentially through the book, temporal guideposts, markers even, the connections to action, are continually drawn. Some of this action – births, marriages and a few deaths for instance – has very little to do with the war. It goes on quite separately, with the war as its backdrop, present but absent. Other forms of action, however, are of critical significance – the decisions of Indians 'fighting' in the British army for instance. And it is the latter type of action that is what is meant by a notion of human agency in the 'making' of history; actions that whilst conditional in themselves in the moments of their happening, can be seen with hindsight as transformative in their effects, and inactions that were equally important in their effects.

One of the effects of developments in geography in the 1990s was to call into question even the possibility of this type of writing, and of the grand narrative that is this type of history. Fortunately though, this type of critique has proved short-lived, and a more provisional, messier sense of the grand historical narrative now receives considerable, even lauded, exposure, and well beyond academic circles (Schama 2001). What this should alert us to is a point that has been much overlooked in recent geographical writing: that human action is highly differentiated in the scope and scale of its effects; and that some actions and some agents matter rather more than others. In the overall scheme of things then it matters not a jot if I let my cat in or make a cup of tea, but what may matter slightly more is where I buy these tea bags from and how such acts of purchase connect to my knowledges about how and where goods like tea are produced. Possibly even, teaching about such things may in turn have effects on the actions of others (though sometimes I doubt it!). Equally though, it is high time indeed that we got back to acknowledging that some people's actions are rather more important that those of others. Tony Blair's globe-trotting exploits for example are newsworthy; those of others I know are not. In itself this is not so much about Tony Blair himself – although it clearly connects with his vision of British foreign policy – but it is rather more about the position that he currently occupies, as British Prime Minister, and the importance others invest in this. To focus once more on the actions of 'the significant', I would argue, is not necessarily to celebrate these individuals (as great men for example – alas, they are almost all still overwhelmingly male!), but it is to give due acknowledgement to the effectivity of positions constituted as powerful.

In what remains of this chapter I want to explore these arguments in a couple of more substantive and expansive illustrative contexts. Both allow me to consider the doubleness of agency/structure: as we will see, they both necessitate that we think in terms of an agency that is more than human, but both too require us to think about the distinctiveness of a differentiated human agency that itself is simultaneously both conditioned and linked to the constitution of events as 'significant', meaningful. Already, both events have been represented by many as transformative in their effects – although we might question to what degree. And both have already been incorporated into grand historical style narratives. It is also to my advantage that – even allowing for academic publication delays – they will be still extremely familiar to most readers. Hence the minimal background.

Killing fields, burning bodies and the re-organization of agricultural matter

> There verdure teems, a growth that for its gladness
> Needs nothing from your tale of seething madness
>
> (Goethe, *Faust II*: 217)

On 15 January 2002 the UK was finally declared 'free' of 'foot-and-mouth' disease. The epidemic – the worst ever globally – lasted over 11 months, to date has cost over £6 billion (Hetherington 2002) and resulted in the premature slaughter of over six million farm animals.

Early April 2001: I am driving up the M6, on my way to a geography fieldclass in Edinburgh. Approaching the Cumbrian town of Penrith I catch my first breath of what is to become a familiar smell – the unmistakable, unforgettable stench, choking and acrid, of burning animal flesh and hair. Then I see them, the pyres and the plumes of smoke: Penrith is literally surrounded, encircled by this activity. From the familiar silhouette of Cross Fell in the East to the Solway Firth to the West – for it is one of those brilliant, translucent days – the eye is filled with bonfire upon bonfire. This is a landscape of mass slaughter, appalling in its scale and extent.

July 2001: Hillwalking in Northern England – one of my passions – is an impossibility. Huge swathes of Cumbria, Northumberland and Durham remain effectively 'closed'. So my partner and I have become a pair of 'cycle nuts' instead. Approaching Canonbie – a village on the Scottish/Cumbria border – we read a huge notice, daubed appropriately in savage, bleeding, red letters: 'Tony Blair: 2000 healthy sheep were MURDERED here, April 4th 2001'. Meaning and blame could hardly be clearer. And there is an eeriness about this village; a silence has descended on its surrounding fields, witness to the dreadful events of that day.

In beginning this section with these personal vignettes, I want to do two things: to emphasize that this is not a straightforward account of the 'foot-and-mouth' epidemic that occurred in England in 2001, and to show that this event impinged on and shaped the actions and everyday lives of many – not just rural communities in the affected areas (primarily the counties of Cumbria, Northumberland and Durham in Northern England, and Devon), where feelings and passions continue understandably to run high, but others who use these areas for leisure and recreational purposes, and those who don't, who were just passing through, travelling the M6. Instead, and to a degree like Law's account of the aeroengine explosion to which I referred earlier, I want to focus on the effectivity of this virus. So, rather than concentrate on the scientific work that went on during this outbreak – on the isolation of its cause and origin (a pig farm west of Newcastle); charting its diffusion through various livestock auction markets in Northern England and Scotland; debating the efficacy or not of a vaccination programme; and the eventual establishment of exclusion zones, the so-called Blue and Red Box zones – I want to highlight two things.

The first is about the virus itself. Watching a recent local television news clip reporting on the re-stocking of a dairy farm near Hawes in Wensleydale, the 'story' flipped back to the day eight months previously when an entire herd, built up by generations of one farming family, was slaughtered. How the farmer talked about this was instructive. Emotional and traumatized, he described how they had done everything humanly possible to preserve this herd: observing stringent disinfectant procedures, isolating themselves, going nowhere, seeing nobody. Yet somehow, 'inevitably' he says, antibodies came to be present in two cows. Antibodies note; not the disease itself. The virus then is seen to have an agency of its own; an agency that transcends human actions and that in its trace – but not its symptoms – determines the fate not just of individual cows but an entire herd, and those on contiguous farms.[1] Occurring at the height of the outbreak, such language and meaning frameworks were perhaps to be anticipated. As many of the 'farmers' diaries' carried by the quality broadsheets at the time showed, feelings of siege yet inexorability were then rife. But somewhat later events reveal exactly the same interpretive frameworks. New Year's Day 2002 saw the 'precautionary' slaughter of 1000 sheep at Donkleywood near Kielder Forest in Northumberland, again because two sheep were found to display antibodies. Once more then, the effectivity of the virus extends well beyond the bodies in which its trace has been detected; and we can see how an area that has been pronounced 'clear' at a national level turns out not to be so. But, another thought: this is the effectivity that we know about, that we can chart and document, where scientific testing procedures mediate the fate of collectivities of animals. But what about some of those more recalcitrant ewes up on various fellsides and hillsides; the ones who always seem to manage to avoid shearing or dipping. Who knows whether they are positive or negative, and what effectivity they might have.

My second point is to emphasize just how much of the activity – that is, intentional human action – following the immediate outbreak and beyond – was about trying to re-establish the established order of things, that is, attempting to prioritize the 'needs' of the farming industry within the overall rural economy and nationally; and trying to utilize procedures for handling the outbreak that had been deployed previously, in 1966. So, actions like the wholesale closure of footpaths countrywide – even in areas well removed from 'hot spots' – and the burning of animals' bodies, as well as the refusal of the National Farmers' Union to agree to a mass vaccination programme can usefully be interpreted as designed to preserve a set of conditions, these being, respectively, the export market for 'British' meat, the authority of MAFF, (the Ministry of Agriculture, Fisheries and Food) to administer and regulate farming practices and the precedence of farmers within rural economies. What happened though, and what makes the course of the 'foot-and-mouth' outbreak so interesting, is that the established order of things continued to be challenged. Matter and boundaries refused to return to previous configurations.

Some of the most graphic illustrations of this occurred in the lanes of Cumbria and Northumberland, where piles of dead animal bodies lay spewing out of farm gates in villages for days on end; bloated and leaking matter from 'private' farmyards onto 'public' roads. Roads onto which lorries, with supposedly sealed units, leaked all manner of fluids as they transported their cargoes to the mass burial pits of Great Orton and Widdrington. The mess that is the mass slaughter of farm animals, animals that are produced for mass, cheap consumption, spilled out from the abattoir and into full public view. Similarly the shortcomings of MAFF procedures, compounded by a lack of available vets, brought about excessive and well-documented backlogs in slaughter programmes. Animals condemned to an earlier than anticipated death remained as potential sources of infectivity to others; present when they should have been absent. At which point the Army was called in, and MAFF's days were numbered. I could add to this the refusal of local residents to be satisfied that mass burial pits, any more than pyres, were not a public health risk; the subsequent re-excavation of the burial pits and the transportation of ash elsewhere; the failure of many local rural businesses, particularly those dependent on the tourist and outdoor leisure industries; the establishment of Defra (the Department for Environment, Food and Rural Affairs); the removal of the then Agriculture Minister, Nick Brown; the continued closure of large swathes of the countryside pending 'cleansing operations' that seem to remain pending; not to mention the plethora of notices that continue to adorn footpaths – red, green and yellow – of varying vintages and contested status. But the point is made: not only did matter refuse to remain in place, but previous boundaries and their relative authorities were, if not always overturned, then at least called into question.

The 'foot-and-mouth' outbreak then remains an event without closure. Its trajectory has left yet unanswered and unresolved a myriad of questions: about the economic future of livestock farming in Northern England; about farming practices and their regulation, particularly about extensive national and international networks of meat production; about the place of farming within rural economies; about the dependency of rural businesses on those who live and work in urban conurbations; about consumer demand for 'British' beef and lamb. All, and more, are important issues. But for me it is another that takes precedence: if the established order, its regulations, practices and series of priorities, has been overturned (possibly a big 'if', but the establishment of Defra certainly enables this possibility), then what we should see is the forging of a new order. And in its making we are likely to see considerable attention being given to the ways in which farming practices produce landscapes; to the ways in which, for example, the English Lake District aesthetic has been constituted by years of over-grazing sheep, and the importance of these close-cropped exposed fells to the tourist and outdoor industries. Pictures of scrubby trees don't quite hold the same attraction as the current Langdale skyline. And who would really prefer thrashing through scrub to walking on springy grass and bare rock? In an inversion of the previous order then, we could see farmers transformed into landscape museum custodians; not taking precedence but acting as curators for the values of often urban-based consumers, who not only eat the food they produce but whose aesthetic sensibilities shape the conditions of food production. The latter, whilst admittedly only one possible outcome, is an example of how events can take on an agency of their own, and how this unfolding both opens up and forecloses particular possibilities and orders. At the same time though, throughout this event conditioned human action and inaction has never been far away. Nevertheless, what may perhaps prove more critical in the longer term is the unintentional, the consequences of actions, which in their effects cannot be controlled – things like consumers switching to poultry purchasing for example, or children's reactions as consumers to walking past dead animal bodies on a daily basis. Or of considerable, prolonged human exposure to the stench of the mass pyres, things that are also a reminder that we do not live in a flattened world of cohabitation, but one where the biographies of certain sorts of animals at least – many of them who just happened to be in the wrong contiguous locations at the wrong time – are terminated by human agency alone, and by human agents whose actions are themselves shaped by the conditions of the farming industry and by meat export markets in particular.

The word, the text and the making of history?

September 11, 2001: As for many, I suspect, this is a day that remains etched in my consciousness. At precisely the moment – 2pm UK time – that we were admiring the perfect Hebridean vista that stretched before us – blue sky, sun glinting on the sea and Ben More completely cloud-free – elsewhere, in yet another sun-filled blue sky, events were unfolding in New York City that were to be both extensive in their spatial effects and intense in their possibilities. For this, indeed, was a moment when many held their breath and waited, for the response they feared from the global hegemonic power. Learning of those events that evening, observing the reactions of American tourists staying in the same small remote Scottish town; listening to some of their conversations and watching the replayed television news images, our day – rightly – came to be overtaken, contaminated by events that had occurred thousands of miles away but which connected us all. Even now I cannot look at pictures of that view without their being superimposed in my eyes' mind by another image; that of a glistening plane arcing round and splicing through the second tower of the World Trade Centre building. Which, hopefully, says a lot about not just time–space compression but about the way in which certain events exert their effect – in the immediacy of their hierarchical relation and beyond the moment of their occurrence.

Much of course has been written about these events and about what has subsequently transpired (see, for example: http://www.zmag.org/ZNET.htm). So much that to add yet another similar account would in all probability be superfluous. Rather, my point here is precisely that so much *has* been written. It is this, and what it might signal, that I explore below.

In the immediate aftermath of that day, in for example the UK newspaper coverage of September 12, it was the visual that prevailed. Page after page of what continues primarily to be a written, if differentiated, media form was dedicated to images: of the moment described above; of the collapse of the towers; of the radically changed Manhattan skyline; of bodies falling through space; and of the firefighters at Ground Zero. Subsequently, much was made in certain critical and Left-leaning circles of the visual symbolism and currency of these images. About, for example, the potency of a civilian aircraft – the means of widespread mass travel and spatial mobility – as a weapon of mass destruction; and about the sophisticated deployment of Western cultural referants and media by those who planned and orchestrated these events – particularly the spectacular conjoining of the cinematic with the corporate and the technical. The constitution of a global television audience for an event that looked like a Hollywood disaster movie, which had at its centrepiece a building that symbolized American corporate power and which, in the final moments of collapse, saw a reversal of those images so often beamed round the world as testimony to American global power – the successful rocket launches from Cape Canaveral. Gradually, however, what we began to see was the appearance of more and more written texts; at first documentary accounts, then more archaeological excavations, analytical explorations from across the political spectrum. I think this is significant. The absence of words initially was not just about the inability of international news teams to get into North America, let alone New York, but about the inabilities of words to capture the moment, the horror of it all. So, what perhaps deserves some reflection is the re-ascendance of the word and why this might have occurred.

What I think this was about, at least initially, was to make sense of these pictures in ways that exceeded the visual; to place these pictures in an order, within appropriate meaning frameworks. Hence the need for factual, chronological accounts, and then the various attempts to account for why this event happened where it did – from the arguments that saw this exclusively in terms of US imperialism, through to those that tried to consider how what 'in Western eyes' appeared irrational and an heinous 'crime against humanity' could be constituted as purposive, intentional and rational by those who perpetrated it. The words then provided a range of interpretive frameworks, and their acts of writing a form of human agency, with the clear intention at least of enabling an understanding, if not always understandings.

But what words also staked out the ground for – and here I am thinking rather more about certain performances in words, specifically the various public speeches and broadcasts of George W Bush and Tony Blair on the one hand and the very different ones of, say, General Musharraf of Pakistan – was a connection between meaning frameworks, human action and the reinstatement of the established world order; notably a highly particular form of largely US military action that saw the repeated bombing of Afghanistan through the autumn and winter of 2001. Action that we should note also resulted in a humanitarian crisis of immense proportions, as Afghan refugees spilled over the borders into Pakistan and Iran, as well as the overthrow of the Taliban regime. Particular, and exclusively Western, interpretive frameworks then were used not just to account for events but to legitimate certain actions, and most definitely not inaction.

It is instructive to think about some of the effects of this action, and to compare what transpired here with my previous example. As numerous political commentaries have remarked, the course of the war in Afghanistan has not simply reinstated the existing world order and the place of the US within this, but threatens to extend this 'solution' (bombing and the replacement of one 'unfavourable' regime with another) elsewhere. As I write, Somalia and Iraq are top of the speculation list. Geopolitically then, established boundaries have been reforged, and speedily. Here then there appears very little of the 'leakiness' that characterized the 'foot-and-mouth' epidemic, and the established order of things seems unthreatened. It is back to 'business as usual' – witness how Afghanistan, Pakistan and India, and to a lesser extent Israel, Palestine, Sharon and Arafat, have disappeared from Western news screens, and how domestic considerations – even trivia – dominate debate, at least in the UK, where currently it is the railway industry (again) and the 'run' on prunes following a Delia Smith television cookery programme that are attracting attention. But is this really business as usual? What about outside 'the West'? Answers here can only be provisional and speculative but the effectivity of actions in Afghanistan has clearly led to heightened tension between Pakistan and India; the Intifada shows little sign of receding; and it is impossible to assess how events will pan out in the Gulf States and Saudi Arabia, where the conjuncture of US military hardware, political regime and a combustible masculine 'Arab street' is a heady cocktail. Matter may have been reinstated in a particular order then, but we need to acknowledge that this is a 'view from somewhere'. Elsewhere, possibilities for leakiness appear acute.

So, what can we say ultimately about this need for so many words? Yes, that they were critical to sustaining a particular course of human action; for the latter certainly required that a majority (albeit a contested one in the UK and continental Europe) subscribed to a particular analysis. But I think too that we can say rather more. At the time of their occurrence, many occupying positions of power – George W. Bush and Tony Blair, for example – described the events of September 11, 2001 as of epochal importance. Metaphors of 'testing' and resolve abounded; 'lines' were 'drawn in the sand'. In short, these events were incorporated immediately within 'history' as of 'historic' importance: think, for example, of the inappropriate parallels drawn with The Crusades or of the continued attempt to link these events to what some regard(ed) as the unfinished business of the Gulf War. It was important then to weave these events into a grand historic and binary narrative, of West against East, Us against Them (Huntington 1996; cf. Chomsky 2001; Said n.d). Why? I think because of their enormity. Incorporating through writing and the analytical frameworks it produces is an act of claiming and ordering; it is a way of corralling an event and of regulating its potential effectivity; a way of allowing us to appear to possess it, rather than be possessed by it. So powerful were these events, for the West, I would suggest, that it was vital for them to be incorporated within a particular version of history, with an immediacy that subsequent events suggests might have been rather less epochal than was claimed at the time.

Concluding remarks

In this chapter I have, in a sense, been exploring the effectivity of a theoretical moment within the agency/structure binary, of the acknowledgement of the agency of the non-human – specifically here things, animals, viruses, aircraft and words – in human geography and its consequences. However, rather than dispense with thinking in terms of an human agency, I have argued that it is vital that we continue to acknowledge the importance of a sense of human agency that still takes intentional, purposive action and its unintentional consequences seriously, and that continues to be connected to the conditions of its production. Admitting the agency of the non-human does not mean that we should abandon the difference that human agency makes. Rather, what is required I think is to bring the effectivity of things and intentional, expressive, human action together, in ways that illuminate what I have called the 'doubleness' of agency/structure; the entwining of effect with purposive act and of networks of effectivity with a conditioned sense of the possible. And it is my hope that the two examples discussed above convey at least something of this. At the same time though, the examples themselves raise two further issues.

One is to show the inseparability of any discussion of agency/structure from an understanding of the event. If we see events as involving the ongoing combination of action and matter in time and space, we can see a particular event or series of events as entailing the bundling up of particular such configurations; that is, as impositions of ordered frameworks of meaning that work to say that some spatially and temporally delineated conjunctures of action and matter actually matter rather more than other arrangements and juxtapositions. Why? For at least two reasons I think. First, because some conjunctures threaten existing arrangements; they potentially destabilize and/or disrupt existing conjunctures of action and matter. Secondly, because the spatial and temporal effects of certain conjunctures mean that their effects extend well beyond the moments and places of their occurrence. My discussion of the 'foot-and-mouth' epidemic illuminated the first of these; September 11, both. As we have seen then, a lot of agency, particularly human agency, is about identifying and determining the potentially meaningful event – that which may or may not have untoward consequences – from that whose effects can be anticipated to be routine and/or mundane, and acting on this however imperfectly or provisionally. Which means that we still require a conceptualization of human agency that recognizes that some human agents have rather more power than others – that they have a greater capacity for action and that their actions will have greater effects; and that we still require a conceptualization of human agency that connects to the sense of histories and geographies in the making.

Secondly, much of the discussion in the previous three sections has shown the inexorable connectivity between the non-human and human. When we talk about the effectivity of things then, we are often talking about how things have an effect upon us and affect us; about, for example, how estate agency works by shaping moralities of practice. But we are also clearly talking too about the ways in which the effectivity of things is shaped by the conditions in which they are located. A point that Marianne de Laet (2000) makes in relation to patents, this is an argument that also applies to the effectivity of the foot-and-mouth virus in Northern England, where the conditions of the meat export industry meant that animal vaccination was not an option in the UK, whereas elsewhere – where 'foot-and-mouth' is endemic – its use is a routine part of farming practice. Agency's conditions then, which includes its geographies, are critical to constituting and delimiting the shape of effectivity.

Finally, I want to end this chapter with a moment that is both a confessional and another plea of sorts. Since this is a text aimed primarily at an undergraduate audience, my expectation is that the response to this chapter of most readers, at least at first, will be to baulk at an argument that talks in terms of the agency of things. Certainly, this was my initial response, and it has been overwhelmingly that of those that I teach. At the same time though, to baulk is to privilege a particular reading of agency, and to dismiss is to re-assert a singular reading located exclusively in human action. As I hope the discussion here has highlighted, thinking about agency currently in human geography necessitates that we think rather more heterogeneously and provisionally about the connections between humans and things, but, or so I would argue, without dispensing with a sense of human agency and its connection to a more conditional sense of structure. If this chapter has had the effect of at least making you the reader think about these things then it – a non-human thing – will have had an effect, and I will have achieved my intention!

Acknowledgement

This chapter is dedicated to Maxwell (1985 – 1 February 2002), who continues to haunt the bin and whose later years were an object lesson in animal agency.

FURTHER READING

Abrams, P. (1983) *Historical Sociology.* Milton Keynes: Open University Press. esp. Preface and Ch. 1

Thompson, E.P. (1963) *The Making of the English Working Class.* Harmondsworth: Penguin. esp. Preface

Two of the most influential earlier accounts of agency/structure, in which agency is understood as human agency.

Thrift, N. (1983) 'On the determination of social action in time and space', *Environment and Planning D: Society and Space* 1: 23–57
A good illustration of the way these ideas about human agency came into human geography.

Latour, B. (1993) *We Have Never Been Modern*. London: Harvester Wheatsheaf. esp. ch. 1
A more recent influential account of agency.

Thrift, N. (2000) 'Afterwords', *Environment and Planning D: Society and Space* 18: 213–55
Compare this account of agency with:

Miller, D. (2001) 'Estate Agency', in D. Miller (ed.), *Home Possessions*. Oxford: Berg

Note

1 It is also worth noting that the agency of the virus extended even beyond its trace, through assumption: 2000 healthy sheep were slaughtered at the end of August 2001 in Northumberland because ulcers were found on the tongue of one animal. These ulcers turned out not to be foot-and-mouth disease but a reaction to the spreading of lime on pasture land (Hetherington 2002).

References

Abrams P (1983) *Historical Sociology*. Milton Keynes: Open Books.

Bingham N (1996) Objections: from technological determinism towards geographies of relations. *Environment and Planning D: Society and Space* 14: 635–57.

Birdwell-Pheasant D and Lawrence-Zúñiga D (eds) (1999) *House Life: Space, Place and Family in Europe*. Oxford: Berg.

Braudel F (1949) *The Mediterranean and the Mediterranean World in the Age of Philip II*. London: Fontana (1973 edition).

Braudel F (1979) *The Structures of Everyday Life: The Limits to the Possible – Volume I of Civilisation and Capitalism 15th – 18th Century*. London: William Collins.

Callon M (1986) Some elements of a sociology of translation: domestication of the scallops and the fishermen of St Brieuc bay. In Law J (ed.), *Power, Action and Belief*. London: Routledge and Kegan Paul. pp. 196–233.

Chomsky N (2001) The world after September 11. http://www.zmag.org/ZNET.htm (accessed 18th January 2002).

Clarke A (2000) Mother swapping: the trafficking of nearly new children's wear. In P Jackson, M Lowe, D Miller and F Mort (eds), *Commercial Cultures: Economies, Practices, Spaces*. Oxford: Berg. pp. 85–100.

Cloke P, Philo C and Sadler D (1991) *Approaching Human Geography*. London: Paul Chapman.

Ghosh A (2000) *The Glass Palace*. London: Harper Collins.

Giddens A (1979) *Central Problems in Social Theory: Action, Structure and Contradiction in Social Analysis*. Basingstoke: Macmillan.

Giddens A (1981) *A Contemporary Critique of Historical Materialism*. Basingstoke: Macmillan.

Giddens A (1984) *The Constitution of Society*. Cambridge: Polity Press.

Gregory D (1981) Human agency and human geography. *Transactions Institute of British Geographers* 6: 1–18.

Gregson N and Crewe L (2002) *Second Hand Worlds*. London: Routledge.

Gregson N, Brooks K and Crewe L (2000) Narratives of consumption and the body in the space of the charity/shop. In P Jackson, M Lowe, D Miller and F Mort (eds), *Commercial Cultures: Economies, Practices and Spaces*. Oxford: Berg, pp. 101–22.

Goethe JW (1959 translation) *Faust*. Harmondsworth: Penguin.

Harvey D (1982) *The Limits to Capital*. Oxford: Blackwell.

Hetherington K (1997) Museum topology and the will to connect. *Journal of Material Culture* 2: 199–218.

Hetherington K and Law J (2000) Guest editorial: After networks. *Environment and Planning D: Society and Space* 18: 127–32.

Hetherington P (2002) Burning issues. *Guardian* 21st February, p 17.

Huntington SP (1996) *The Clash of Civilizations and the Remaking of World Order*. New York: Simon and Schuster.

de Laet M (2000) Patents, travel, space: ethnographic encounters with objects in transit. *Environment and Planning D: Society and Space* 18: 149–68.

Latour B (1987) *Science in Action*. Cambridge MA: Harvard University Press.

Latour B (1993) *We Have Never Been Modern*. London: Harvester Wheatsheaf.

Law J (2000) Transitivities. *Environment and Planning D: Society and Space* 18: 133–48.

Law J and Hassard J (eds) (1999) *Actor-network Theory and After*. Oxford: Blackwell.

Law J and Mol A (eds) (2001) *Complexities in Science, Technology and Medicine*. Durham, NC: Duke University Press.

Le Roy Ladurie E (1978) *Montaillou: Cathars and Catholics in a French village, 1294–1324*. London: Scholar Press.

Lee N and Brown S (1994) Otherness and the actor-network: the undiscovered continent. *American Behavioral Scientist* 37: 772–90.

Ley D and Samuels MS (eds) (1978) *Humanistic Geography: Prospects and Problems*. London: Croom Helm.

Livingstone D (1992) *The Geographical Tradition*. Oxford: Blackwell.

Massey D (1979) In what sense a regional problem? *Regional Studies* 13: 233–43.

Massey D (1984) *Spatial Divisions of Labour*. London: Macmillan.

Mol A and Law J (1994) Regions, networks and fluids: anaemia and social topology. *Social Studies of Science* 26: 641–71.

Miller D (1987) *Material Culture and Mass Consumption*. Oxford: Blackwell.

Miller D (ed.) (2001) *Home Possessions: Material Culture Behind Closed Doors*. Oxford: Berg.

Miller D and Slater D (2000) *The Internet: An Ethnographic Approach*. Oxford: Berg.

Miller D, Jackson P, Thrift N, Holbrook B and Rowlands M (1998) *Shopping, Place and Identity*. London: Routledge.

Murdoch J (1997) Inhuman/nonhuman/human: actor-network theory and the potential for a dualistic and symmetrical perspective on nature and society. *Environment and Planning D: Society and Space* 15: 731–56.

Murdoch J (1998) The spaces of actor-network theory. *Geoforum* 29: 357–74.

Said E (n.d) The clash of ignorance. http://www.zmag.org/ZNET.htm (accessed 18th January 2002).

Schama S (2001) *A History of Britain, Volume II: The Wars of the British 1603–1776*. London: Hyperion.

Thompson EP (1963) *The Making of the English Working Class*. London: Victor Gollancz.

Thompson EP (1979) *The Poverty of Theory*. London: Merlin Press.

Thrift N (1983) On the determination of social action in time and space. *Environment and Planning D: Society and Space* 1: 23–57.

Thrift N (2000) Afterwords. *Environment and Planning D: Society and Space* 18: 213–55.

Tuan Y-F (1976) Humanistic geography. *Annals of the American Association of Geographers* 66: 266–76.

Wallerstein I (1974) *The Modern World System, Volume I*. London: Academic Press.

Whatmore S (1997) Dissecting the autonomous self: hybrid cartographies for a relational ethics. *Environment and Planning D: Society and Space* 15: 37–53.

Whatmore S and Thorne L (2000) Elephants on the move: spatial formations of wildlife exchange. *Environment and Planning D: Society and Space* 18: 185–204.

3 State : Society

Joe Painter

In this chapter I focus on the state/society pairing, which has been an important one in the development of geographical thought, dating back at least to the work of the nineteenth-century German geographer Friedrich Ratzel (1844–1904). Moreover, state/society relations have a profound influence on the human geography of the world, shaping spatial patterns, changing places and animating landscapes. And yet to use the phrase 'state/society relations' at all, is to accept implicitly that 'state' and 'society' are separate, distinct phenomena between which there can be defined 'relations'. In other words, by thinking of 'the state' as something that influences 'society' (or vice versa) as if the state were something outside of society, we are already in danger of accepting at face value the very idea that this chapter sets out to question. Before that questioning starts in earnest, though, it will be useful to consider whether the state/society pairing should be thought of as a binary relationship in the same sense as some of the pairings examined in other chapters.

The idea of a binary relationship implies a categorical distinction of the form either/or that divides all cases into two groups (men/women; white/black; young/old etc). Binary relationships can also be expressed in the form 'A or not-A'. This version entails a strong power relationship between the two categories. Whereas either/or might be seen as expressing difference with equality, in 'A or not-A', the second category is defined both entirely by reference to the first, and in terms of a lack of the qualities of the first (rather than in terms of its own positive features). When binary relationships are understood as 'A/not-A', women are defined as 'not men', black is defined as 'not white' and old is defined as 'not young'. These kinds of pairings tend to imply that the second category ('not-A') is inferior and subordinate to the first, and distinguished not by its own substantive content and value, but only by its difference from the first category and the absence of those qualities that define 'A'.

Recent critiques of binary thinking in the humanities and social sciences emphasize this power relation and the implied subordination it involves. Some feminist writers, for example, have argued against the use of binary categories such as men/women and masculine/feminine on the grounds that they define women/feminine exclusively in relation to men/masculine and also as inferior.

The use of these categories disguises the differences that exist between women (and men) and underwrite essentialist definitions of masculinity and femininity. They deny the possibility of gender identities that are not exclusively masculine or feminine and are unable to accommodate the existence of human physiologies that are neither entirely male nor entirely female.

Binary thinking has undoubtedly been influential in the development of geographic thought, and the critiques of it are as important to geography as they are to the rest of the humanities and social sciences. Many of the chapters of this book are devoted to assessing the influence of binary thinking in geography and elaborating those critiques. Such an approach, however, cannot be applied straightforwardly to the subject of this chapter: the pairing of state/society. This is because the state/society pairing cannot really be presented as a binary relation of the form 'A/not-A' in the way that male/female and white/black can. Western social thought has had an unfortunate tendency to equate 'female' with 'not male'; but it has rarely defined 'state' as 'not society'. The definition of 'society' as 'not state' is perhaps slightly more common, but overall the categories of 'state' and 'society' have not typically been treated as a binary pair of the form 'A/not-A'. Moreover, in geographic thought 'state' and 'society' have often been the focus of attention of different subdisciplines. The state has been the concern above all of political geography, whereas the concept of society has been particularly important in social geography and development geography. This means that the geographic literatures on state and society have often spoken past each other and thus in geography the concepts of state and society have not been 'co-constitutive'; that is, they have not always been defined in relation to each other, but often as separate objects of study, each in its own right.

Feminist geographers have been able to use the analysis of binary thinking (among other things) to develop a thoroughgoing critique of the masculinism of geographic thought in general. Similarly post-colonial geographers have been able to launch an assault on geography's Eurocentrism by deconstructing binary pairings such 'the West and the rest' or 'Europe and its Other(s)'. By contrast, ideas about the 'state/society' pairing have not shaped geographic thought in general to quite the same (pernicious) extent as has binary thinking about gender and 'race'. In other words, a critical re-appraisal of geographical thinking about the state/society pairing may not have the same far-reaching consequences for geography in general as the critical re-appraisals developed by feminist and post-colonial writers.

If 'state/society' is not a binary in quite the same sense as some of the other pairings, then our critique of it is also likely to involve a different approach. In what follows, therefore, I will consider some of the ways in which the relationship between state and society has been understood in geography, focusing among other things on the question of *separation* (the assumption that state and society are separate entities that interact). The chapter will also examine recent geographical ideas – and draw on debates in the wider social sciences – that call that separateness into question and seek to re-theorize the 'state/society' pairing and its geographies.

Geographical approaches to the state

In some respects the history of geography's engagement with the state is a strange one. In the nineteenth and early twentieth century the state was at the core of geographers' concerns. Throughout the middle years of the twentieth century, up to and including the period of the so-called quantitative revolution in Anglo-American geography in the 1960s and 1970s, the state was largely neglected as an object of geographical enquiry, or treated in a rather limited or simplistic fashion. Then, at the very end of the 1970s, and especially in the 1980s the state was rediscovered as part of a major renewal of anglophone political geography. Today, geographers studying all kinds of things from social exclusion to economic restructuring and from the historical geographies of madness to post-coloniality are interested in the state, and their insights have contributed to many new ways of understanding the formation, functions and spatialities of states. In what follows I shall limit the discussion to the modern academic discipline of geography, originating in the middle years of the nineteenth century, and primarily, though not exclusively, to the anglophone literature. (For a more detailed discussion of the history of political geography see Agnew 2002).

The state as organism

The work of the Friedrich Ratzel was enormously influential in the development of modern geography as a whole, but it also provides one of the earliest geographical discussions of the state. Indeed, the concept of the state was central to much of Ratzel's work. For Ratzel the state was a living organic entity that expressed the unity of land and people. Thus, in 'The laws of the spatial growth of states', first published in 1896, he wrote:

> States are dependent both in their size and their form upon their inhabitants, i.e. they take on the mobility of their populations, as it is particularly expressed in the phenomena of their growth and decline. Some number of people are joined to the area of the state. These live on its soil, draw their sustenance from it, and are otherwise attached to it by spiritual relationships. *Together with this piece of earth they form the state.* For political geography each people, located on its essentially fixed area, represents a living body which has extended itself over a part of the earth and has differentiated itself either from other bodies which have similarly expanded by boundaries or by [*sic*] empty space. (Ratzel 1969: 18; emphasis added)

The phrase 'together with this piece of earth they [i.e. the people] form the state' strongly suggests that for Ratzel the relationship between state and people (it is notable that he does not use the word 'society' here) was a unity, rather than a binary. In other words 'the people' were integral to the state, not separate from it. In contrast to more recent definitions of the state as a set of differentiated institutions (see below), Ratzel's sees the state as made up of the population. This way of conceiving the state owes much to Hegel:

> For Hegel, the State is a realization of the Spirit. Thus, the latter does not reveal itself to the peoples, but it is realized in them – for them, when they become conscious of it – by their own labour, by their conscious and reasoned action. (Raffestin et al. 1995: 85; my translation)

Although Hegel saw the state in organic terms, he also broke with early contract theory, which understood the state and civilized society to be the same thing, in contrast to the *un*civilized state of nature. In *Philosophy of Right*, Hegel distinguished conceptually between The Family, The State, and Civil Society, though he includes in 'Civil Society' the police and the administration of justice which are more usually thought of today as part of the state.

It is easy to see how arguments about the essential oneness of the state, the national territory, and a culturally distinct people might be used to underpin exclusionary forms of nationalism and even fascism. In addition, Ratzel argued that as organic entities, states have an inevitable tendency to grow geographically. Such a view was all too easily used to provide supposedly 'scientific' justifications for territorial ambitions and expansionist policies of the German state. Ratzel died some 30 years before the rise of Nazism in Germany, and while writers such as Raffestin counsel against drawing too direct a connection between the work of Ratzel and that of the proponents of *Geopolitik* (which informed Nazi policy), he also affirms that there is an undeniable continuity between Ratzel's ideas and those of writers such as Rudolf Kjellen and Karl Haushofer (Raffestin et al. 1995: 121).

The state and conservative political geography

It was partly because of the disreputable association of geopolitics with Nazism that political geography declined in strength and popularity after World War II (Agnew 2002: 94). For 20 or 30 years political geography was largely evacuated of any concern with politics. In the words of Ron Johnston, writing as late as 1980,

> political geography is weakly developed within the current corpus of human geographical work because its practitioners ignore the real content of politics, which is the exercise of power by and through the state. (Johnston 1980: 439)

This state-centric view of political geography has in turn itself been challenged by more recent work that stresses the ways in which 'the real content of politics' is not exhausted by the power relations of state. None the less, Johnston's wider point that politics is about social conflicts and power, and that the subdiscipline of political geography had been enfeebled by neglecting them was undoubtedly well founded.

While geographers in the 1950s and 1960s may have avoided tackling issues of power and social conflict, it is not quite true that they ignored the state. Richard Hartshorne (1899–1992), for example, the doyen of mid-century American geography, initiated his discussion of 'the functional approach in political geography' by proposing 'to consider the central problems of political geography in terms of the functions of state-areas' (Hartshorne 1950: 104). The suffix '-areas' is indicative of Hartshorne's attempt here to align political geography with his over-arching understanding of the wider discipline as the study of 'areal differentiation' (Hartshorne 1939). For Hartshorne, geography was principally about the study of the character of areas and thus the main concern of political geography was the study of political areas, foremost among which are state territories. At the core of Hartshorne's argument, therefore, was a concern with the territorial integrity of the state – with the 'centrifugal' forces tending to fragment the unity of the state and with the 'centripetal' forces tending to reinforce that unity. This focus on formal territorial unity and the spatial organization of the state takes precedence over any concern with the nature of the relationship between state and society. Although Hartshorne is careful to reject the Ratzelian argument that the state is an organism, he nevertheless uncritically cites Ratzel's definition of the state as 'a section of land and a section of humanity organized as a single unit in terms of a particular, distinctive idea' (Hartshorne 1950: 110). For Hartshorne, the nature of the relationship between 'state' and 'society' is captured mainly in his highly conservative conceptualization of the nation as the mechanism binding a people to 'their' state (Hartshorne 1950: 113–14). Thus, Hartshorne, unlike Ratzel, seems to regard 'state' and 'nation' as distinct and potentially separate phenomena – functional correspondence between state and nation might be highly desirable, but is not inevitable. This suggests that in addition to the dominant spatial or areal definition of the state Hartshorne is also working with an institutional definition. Although its nature is not discussed in any detail, there is an implication that the state must be understood at least partly as a differentiated institutional apparatus, rather than as a mystical union of people, land and spirit.

Hartshorne's static and apolitical conception of the state set the dominant tone until the early 1980s. While John Agnew sees a more dynamic theory of the state in the work of Jean Gottmann (Gottmann 1952), Peter Taylor and Colin Flint are more sceptical, arguing that Gottmann's work shares Hartshorne's conservatism (Taylor and Flint 2000). In a recent commentary, Taylor (2003) goes further still, denouncing this '"non-social" state theory' as one of the dominant strands in the creation of 'a most unsuccessful conservative sub-discipline [of political geography]'. He continues: 'for most of the twentieth century conservative political geography was under-researched, pedagogically incoherent, perennially in crisis, and, not surprisingly, widely ignored by the rest of human geography' (Taylor 2003: 48).

The emergence of 'spatial science' in the 1960s (which was critical of the Hartshornian orthodoxy) did not significantly improve matters. While electoral geography benefited from the development of more sophisticated quantitative techniques, other aspects of political geography, including studies of the state, could not easily be assimilated to the spatial scientific project. Spatial science tended to focus on the spatial attributes of phenomena in isolation or abstraction from their social and political content. In the case of the state this approach led to a concern with the 'purely' spatial aspects of states (for example size and shape) or, at best, their territorial character. Ed Soja, for example, examined the territorial integration of the state in East Africa through a quantitative analysis of communications flows (Soja 1968). More recent social theoretic approaches also stress territoriality of course. Michael Mann famously argued that the state is defined in part by its territorially centralized character (Mann 1984). Moreover, for Mann, territoriality is a source of *autonomous* power for the state. In other words, the state is not exclusively dependent on power derived from the economy or society (as in some versions of Marxism, for example). However, that does not mean that the territoriality and spatiality of the state can be understood without reference to social and political processes in the way implied by spatial scientific approaches to the state.

As this brief survey suggests, geographers studying the state in the third quarter of the twentieth century did not theorize the state in a binary relationship to society. In general, the relationship between the state and society was either ignored entirely in favour of a focus on the narrowly spatial aspects of state form, or was taken for granted and left unexamined. Soja's bland and uncritical assertion, based on a citation of Hartshorne, that 'a primary function of any politically organized area is to integrate effectively its territorial components – to create a community of interests which accommodates innovation, sustains development, and promotes the general welfare of its adherents' (Soja 1968: 39) is typical and open to challenge on both empirical and theoretical grounds.

Geography, political economy and the state

By the 1970s, therefore, Anglo-American geography's understanding of the state was ripe for a wholesale reinvention, and that is exactly what it got. One of the earliest contributors to this re-thinking was David Harvey, who, as part of his wider engagement with Marxist political economy, published a paper on 'The Marxian theory of the state' in *Antipode* (Harvey 1976). The paper does not discuss the *geography* of the state in so many words. As its title suggests, it is more an exegesis of (then) current Marxian state theory than a geographical re-working of it. None the less it does contain several features that are to be found in much subsequent geographical scholarship on the state.

First, the arguments of the paper both derive from, and act as a commentary on, the literature on Marxist state theory published (or translated into English) during the late 1960s and early 1970s. In particular, Harvey's paper provides one of the earliest discussions in the geographical literature of work on the state by Ralph Miliband, Antonio Gramsci, Bertell Ollman and Nicos Poulantzas as well as Marx and Engels. Secondly, Harvey draws attention to the ways in which the state sustains and guarantees many of the fundamental relationships of capitalism. These include a system of property rights, regulation of the supply of money and credit and the provision of certain public goods 'which are necessary prerequisites for capitalist production and exchange but which no individual capitalist would find it possible to provide at a profit' (Harvey 1976: 84). Furthermore, although the state represents the interests of capital, it can act as an arbiter (though not always a neutral one) between different fractions of capital. Thirdly, Harvey points out the fragmentation of the state into different institutions. This means that the state is understood as a differentiated institutional apparatus, and not a homogeneous monolith. At the same time, Harvey is insistent that the state cannot be defined as the sum of its institutions. Rather 'the state should in fact be viewed, like capital, as a *relation* or as a *process*' (1976: 87; original emphasis). Fourthly, although most of his paper is concerned with general theoretical principles, Harvey is clear that abstract concept of 'the state' should not be reified in the study of particular social contexts:

> the state is not an appropriate category for describing the actual processes whereby power is exercised. To appeal to the category 'the state' as a 'moving force' in the course of concrete historical analysis is, in short, to engage in a mystification. (1976: 87)

This insight also focuses attention on the process of state formation. Thus 'the bourgeois state did not arise as some automatic reflection of the growth of capitalist social relations. State institutions had to be painfully constructed' (Harvey 1976: 87).

What can be inferred from Harvey's paper about the nature of state/society relations? In keeping with much Marxist work on the state, Harvey understands the state as part of society, not separate from it. States are integral to the workings of capitalist societies which would be unable to function without particular functions provided by the state. Furthermore, while the institutional aspects of the state are important, the state cannot be conceived as a set of institutions separate from society. Capitalist social relations run through the state, conditioning, though not wholly determining, its form and function. Harvey's paper thus provides a new way (for geography) of understanding the state as a differentiated aspect of society.

The paper closes with three 'unresolved questions' for future research. As these have animated much geographical study of the state in the years since Harvey's paper was written (whether directly or not), it is worth setting them out in full:

1 To what degree do the various aspects and instrumentalities of state power yield to the state a relatively autonomous function in relationship to the path of capitalist development, and to what degree can state functionaries act as purely neutral or even self-serving arbiters in class and intra-class conflict? These questions have been in the forefront of much of Poulantzas' recent work.
2 To what degree can the capitalist state vary its forms and structures to give the appearance of quite substantial differentiation amongst the capitalist nations, while fulfilling the basic function of sustaining a capitalist society and ensuring the reproduction of that society? In other words, what variety of institutions is possible given the assumption of a basic underlying purpose to state action.
3 Which structures and functions within the state are 'organic' to the capitalist mode of production and therefore basic to the survival of capitalist social formations and which are, in Gramsci's phrase, purely conjunctural? (Harvey 1976: 89)

Political economy approaches, and particularly their historical materialist variants, have dominated geographical theorizing about the state since the mid-1970s, although liberal and public choice paradigms have also been developed by some writers. Among geographers, Michael Dear, Gordon Clark and Ron Johnston all worked on the three issues raised at the end of Harvey's paper (Clark 1981; Clark and Dear 1981; Clark and Dear 1984; Dear 1981a; Dear 1981b; Dear and Clark 1978; Johnston 1980; Johnston 1982).

In their *State Apparatus* (1984), Clark and Dear, like Harvey, understand the state as intimately entwined with the social relations of capitalism. Capitalism, they write, 'is as much a political system as it is economic' and 'state intervention in sociospatial processes is a social event, embedded within society and deriving its logic from society as a whole' (1984: 2). 'In our terms' they stress 'the theory of the state and sociospatial processes cannot be divorced from some wider concept of society itself' (1984: 3). The wider concept that they adopt is historical materialism. However, in a passage that bears directly on the issue of the relationship between state and society, they differentiate their approach from some other versions of Marxism:

> [O]ur theoretical stance is *state-centred*, in that we begin our analysis from the viewpoint of the state as it is embedded within the structural relations of capitalism. This mode of analysis stands in sharp contrast to the more traditional theories of the state, be they marxist or liberal, which are *society-centred*. In these theories, society is the initial and principal object of inquiry, and the state's role is derived from the ensemble of class or individual relations. (Clark and Dear 1984: 9; original emphasis)

For Clark and Dear, the state-centred approach that they favour 'focuses upon the actual behaviour of the state as an institution' (1984: 9), whereas in the society-centred approach 'the analytical method focuses upon the social obligations between society's members, and hence derives the necessity of the state, and its apparatus, from social relationships' (1984: 10). They argue that many liberal formulations are society-centred, in that they see the state as a necessary means for adjudicating between competing individual claims on resources. Marxist work on the state has also been largely society-centred, they contend. The ideas of Poulantzas (1973) and Miliband (1973) are often seen as opposed to one another, but Clark and Dear suggest that they share a common society-centred perspective in which the nature of the state is 'derived from the logic of capitalism itself' (1984: 11). Thus, according to Clark and Dear, liberal and Marxist theories 'begin with social units, individuals or classes, then build society around either shared or antagonistic preferences, and finally derive the necessary collective functions for social continuity. The state is a product of these logical derivations' (1984: 11).

Although Clark and Dear praise the work of Poulantzas and Miliband, in the end they find society-centred theories inadequate because they are not able to account for the 'non-necessary' features of the state – those functions that the state undertakes not because it logically has to for capitalism to survive, but because it has developed 'within a particular historical and geographical context' (Clark and Dear 1984: 12). Thus,

> [t]he issue is then to consider the capitalist state in its entirety, not only its logical derivation from class relationships, and to consider also its particular institutional form, functions and apparatus, and not to regard it merely as a hegemonic integrated institution. [... W]e do not deny the utility of society-centered theory, but rather seek to understand the actions of the capitalist state in specific contexts, a goal which also requires a state-centered methodology. (Clark and Dear 1984: 11)

This methodology leads Clark and Dear to emphasize the *relative autonomy* of the state *vis-à-vis* society, focusing on the state as a set of institutions and practices. This analysis takes up the remainder of their book and considers issues of political language, law, local government, democracy, legitimacy and justice.

Rethinking the state/society distinction

State-centred perspectives, of which Clark and Dear's work is an example, have not been without their critics, however. Writing of developments in American political science, Timothy Mitchell (1991) notes two divergent responses to the difficulty of defining the state. The first response is to abandon the concept of the state altogether in favour of the concept of the political system. By contrast,

> [t]he second response, since the late 1970s, has been to 'bring the state back in' (Evans, Rueschemeyer, and Skocpol 1985). The new work on the state has defined the term in a variety of ways, most of which take it to be not just distinguishable from society, but partially or wholly autonomous from it. (Mitchell 1991: 77)

Mitchell suggests a third approach in which

> the elusiveness of the state–society boundary needs to be taken seriously, not as a problem of conceptual precision but as a clue to the nature of the phenomenon. Rather than searching for a definition that will fix the boundary, we need to examine the detailed political processes through which the uncertain yet powerful distinction between state and society is produced. The distinction must be taken *not as the boundary between two discrete entities but as a line drawn internally within the network of institutional mechanisms through which a social and political order is maintained.* (Mitchell 1991: 78; emphasis added)

This insight helps to explain why it is so difficult to define the state as comprising a particular group of institutions. As I have argued elsewhere (Painter 2000) it is in practice often very difficult to determine whether a particular institution is part of the state or not. The contemporary political scene abounds with institutions that are neither wholly part of the state, nor wholly outside it, such as regulated utilities, private prisons and church schools. Even non-governmental organizations, which by their very name declare their separateness from the state, are often today so intimately entwined with processes of governance as to call their 'non-governmental' status very much into question. Such difficulties vindicate Harvey's insistence that the state should be understood as a process or a relation.

Mitchell's concerns about the problematic nature of the state/society distinction have been endorsed and developed by Bob Jessop, a social and political theorist whose work has been particularly influential in human geography. According to Jessop, neo-statism (that is, state-centred theory):

> assumes there are clear and unambiguous boundaries between state apparatus and society, state managers and social forces, and state power and

> societal power. It implies that the state [...] and society are mutually exclusive and self-determining, each can be studied in isolation, and the resulting analyses added together to provide a complete account. This reifies and renders absolute what are really emergent, partial, unstable, and variable distinctions. It rules out hybrid logics such as corporatism or policy networks; divisions among state managers due to ties between state organs and other social spheres; and many other forms of overlap between state and society [...]. If this assumption is rejected, however, the distinction between state-centred and so-called society-centred approaches dissolves. (Jessop 2001: 155)

It should be emphasized that the question of whether the state is part of society or separate from it is not merely a matter of semantics. That is, it is not just an issue of extending the scope of the *definition* of the word 'society' to include the institutions that are conventionally thought to make up the state. Rather, as writers like Mitchell and Jessop point out, what is involved is a change in the way in which the state is conceptualized, thereby challenging the reification of the state and emphasizing the relational and processual nature of state formation (Jessop 2002: 37).

Jessop himself draws on (among other things) the neo-Marxism of Antonio Gramsci and Nicos Poulantzas to develop a strategic-relational theory of the state. Starting from Poulantzas's (1978) insight that the state is a social relation, Jessop defines that state as

> a relatively unified ensemble of socially embedded, socially regularized, and strategically selective institutions, organizations, social forces and activities organized around (or at least involved in) making collectively binding decisions for an imagined political community. (Jessop 2002: 40)

The state is thus understood as both the product of the interaction of social forces and competing interests and the arena within or through which those forces and interests interact. However, the state is not simply the expression of social relations constituted somewhere else; rather, its institutional form has significant effects on the nature of those relations. The strategies of social and political actors, including (but not only) state managers, are affected by the state differentially, so that the state is not only strategic, but strategically selective – put simply, some strategies do better than others. As Jessop puts it, 'state power reflects the prevailing balance of forces as this is institutionally mediated through the state apparatus with its structurally inscribed strategic selectivity' (2002: 40).

Strategic relational state theory does away with the idea of separate spheres of 'state' and 'society' that are self-contained and interact. It emphasizes that social relations cannot be straightforwardly distinguished from state institutions, rather the two are mutually constitutive, albeit in selective ways.

During the 1990s the strategic relational approach became influential within human geography, particularly in studies of neo-liberalism and the rise of the 'workfare state' and of the changing scales of governance and regulation under globalization (for example, Jones 1997; Brenner 1998; MacLeod and Goodwin 1999; Jessop 2000; Brenner 2001; Peck 2001a, 2001b; Painter 2002b; Peck 2002).

Beyond 'embedded statism'

Questions of scale have also been at the heart of debates about another kind of state-centrism, which also have a direct bearing on the issue of the state/society relationship. In this case, though, the central issue is not how to define the state, but how to define 'society'. The word 'society' is such a common term in everyday discourse that it is easy to assume that its meaning is clear and unproblematic, but it is nothing of the sort. Moreover, its problematic character is rarely examined even in academic social science. The unspoken assumption behind most uses of the term 'society' is that it shares its geographical boundaries with the state. In other words, just as the world is divided into many states ('Britain', 'France', 'China' and so on), so it is assumed that it is also divided into an equal number of national societies ('British society', 'French society', 'Chinese society' and so on) (Taylor and Flint 2000: 6). Indeed this territorial equation of 'society' with 'state' recurs throughout the social sciences. Thus, the unit of analysis in sociology has traditionally been 'society' understood as consisting of the national population and in (macro-)economics it has been national economies ('the British economy', 'the French economy', 'the Chinese economy'). Nation-state thinking is so dominant that this tendency to divide social processes up along state-territorial lines is mostly taken for granted and usually goes unremarked (Häkli 2001). This has often been true of radical and Marxist accounts as well as liberal and conservative ones. For example, David Harvey's account of the capitalist state discussed above refers to 'capitalist social formations' (Harvey 1976: 86) and 'actual capitalist societies' (Harvey 1976: 87), in both cases in the plural.

The world-systems approach developed by Immanuel Wallerstein is an exception to this rule (Wallerstein 1974, 1976, 1979, 1983). World-systems theory was introduced into geography by Peter Taylor in the early 1980s and has formed the basis of a heterodox strand of political geography for the past 20 years (Taylor 1981). Wallerstein argues that there is a single 'world-system' which forms the object of social enquiry and thus world-systems theory adopts a 'single-society assumption' in contrast with the 'multiple-society assumption' of conventional social science (Taylor and Flint 2000: 6). The multiple-society assumption is at the heart of what Taylor calls 'embedded statism' (Taylor 1996). This refers to the way in which the spatial organization of the world into apparently natural sovereign states structures how the world is understood by social scientists. As Taylor puts it:

> the state-centric nature of social science faithfully reflected the power containers that dominated the social world it was studying. Its failure to problematize this spatial structure can be explained by the naturalization of nation-state territories. (1996: 1920)

Embedded statism is a problem for a number of reasons. Taylor's main concern is that it leads social scientists to mistake their objects of analysis and fail to recognize the role of world-systemic processes in shaping national and local circumstances. In relation to the focus of this chapter, it disrupts the state/society pairing in a particular way. The 'separate spheres' view (that state and society are distinct entities that interact) assumes that the spheres occupy the same geographical space and have the same boundaries (as in the assumption the British state governs British society and the British economy). However, Taylor's critique shows that this characterization of state/society relations is fundamentally flawed. To be sure, the British state is real enough and occupies a demarcated territory. Indeed the territoriality of the modern state is such that the Concise Oxford Dictionary offers 'a state' as one possible definition of the word 'country'. The concepts of 'British society' and 'British economy' however are much more problematic. It is not necessary to accept world-systems theory *in toto* to agree that the idea of a national society or a national economy represents an attempt to place territorial boundaries around processes and relations that have much broader spatial scope.

The risk in Taylor's approach is that it may underplay the extent to which the form and function of territorial states generate territorial differences in economic, social and cultural structures and processes. While the ideas of 'British (or French or Chinese) society' or 'the British (or French or Chinese) economy' may be imaginary, they are also imagined (not least by state actors) and that leads to a welter of state activities that seek to make them real. It may be impossible to draw a sharp territorial boundary between 'British society' and 'French society', but social processes operate (somewhat) differently in Britain and France and there are real differences between social structures and social institutions in the two countries. This does not mean that Taylor's 'single-society assumption' is wrong, merely that it is also important to remember that the single (world) society has a geography that includes territorial differences that are in part generated by and also reflect the world political map of nation-states.

To insist on the constitutive role of state processes is not necessarily to return to the kind of state-centrism criticized by Mitchell and Jessop. Jessop himself, and geographers working with his ideas, have combined a rejection of calls for state-centred theory with an emphasis on the state's role in the production and transformation of social and economic life. One of the central aims of regulation theory, of which Jessop is a leading exponent, is to disclose the nature and causes of national differences in socio-economic relations (Jessop 1997; for a critique see Cox 2002). The recent work on the production of scale that was mentioned above also focuses on the ways in which the state's 'spatial selectivity' (Jones 1997) structures the scaling and re-scaling of social and economic life.

State/society in contemporary human geography

Until comparatively recently geographical work on the state was largely confined to those with a specialist interest in political geography. In many cases 'the state' was viewed as the political geographer's central object of study (Johnston 1982) in the same way that 'the economy' was held to be the focus of economic geography. Elsewhere I have criticized traditional political geography's obsession with formal political institutions (Painter 1995) and with the state (Painter 2003) to the neglect of informal relations of power and post-sovereign political forms. That does not mean, of course, that the state is somehow unimportant. Indeed, it might be argued that it is too important to be left to political geographers and in recent years it has not been. Social geographers, economic geographers, historical geographers and development geographers and others have all been turning their attention to the state to a greater extent than ever before. This has brought a welcome empirical and theoretical pluralism to geographical studies of the state and the 'state/society' coupling.

A good example of this greater theoretical pluralism is the growing importance of feminist perspectives. In a study of the treatment of the state by Latin Americanist geographers, Altha Cravey suggests that

> a feminist perspective on the state offers powerful insights into Latin American politics, society and social change. Understanding the state as capitalist and patriarchal enables a nuanced mapping of social dynamics. That is, a sensitivity to shifting power structures within society and within the state can help researchers transcend static and bounded representations of formal politics. Realignments of power within and beyond the state are therefore more easily anticipated and appreciated. Social scientists might not have been as surprised at the strength of women's movements which confronted authoritarian regimes in the 1980s if the gendered nature of power had been understood. Women cultivated and maintained international, regional and local networks which influenced the institutionalized forces of the states they confronted. (1998: 524)

Whereas her emphasis on the gendering of the state is apposite, Cravey's assertion that states are responsible for 'bounding territory and enclosing societies' (1998: 524) seems to run against Taylor's 'single-society assumption'.

On the other hand, her substantive discussion of the nature of state strategy does focus on similar issues to those addressed by the strategic relational approach, including the 'gendered division of labor within the state apparatus; the gendered structure of power in the state apparatus; the gendered structure of cathexis (or emotional attachments); and the interplay between social movements, state policies and outcomes' (Cravey 1998: 526).

Feminist ideas have also been important in the recent growth in geographical studies of citizenship – another issue where the conventional clear distinction between state and society does not hold. Citizenship provides a good illustration of many of the arguments discussed above because it combines formal political rights derived from the state with socio-cultural questions of identification and recognition. Luin Goldring's study of the geography of citizenship across the US–Mexico border suggests that Mexican women, while frequently excluded from the formal categories of citizenship, are able to exercise 'substantive social citizenship' across the border in the US (Goldring 2001). It is clear that citizenship relations are neither completely 'part of the state' nor 'part of society', but the arguments of this chapter suggest that it would also be incorrect to suggest that they are formed at the 'interface' between state and society as this still implies that state and society are discrete entities. Instead, citizenship is an arena in which the neat distinction between state and society breaks down, giving rein to what Jürgen Habermas calls the colonization of the lifeworld by the system (Habermas 1987). The state becomes ever more bound up with the practices of everyday life making it impossible to distinguish between aspects of life that are unambiguously within the state and others that are unambiguously outside.

Citizenship rights and ever-increasing state regulation are of vital concern to one particularly vulnerable social group: people diagnosed as mentally ill. As Chris Philo's exhaustive study of the historical geography of the 'mad-business' in England and Wales reveals (Philo 2004), state actors and institutions have long had an interest in the diagnosis, treatment and containment of mental illness. This activity extends beyond anything that might be called 'state intervention' (which implies action taken from outside) and can more accurately be characterized as a process of active constitution in which medical practitioners (today of course mostly employed by the state), the judicial system, government departments and local officials are engaged in a constant, if geographically differentiated (Philo 2004), process of describing, defining and organizing the phenomenon of mental illness and the worlds of those living with it.

Philo's work draws extensively on Michel Foucault and it is also from Foucault that recent geographical writing on governmentality draws its inspiration (examples include Luke 1996; Murdoch and Ward 1997; Braun 2000; MacKinnon 2000; Moon and Brown 2000). The concept of 'governmentality' refers to the ways in which the state constitutes its own objects of governance (Foucault 1979; Burchell et al. 1991; Dean 1999). In other words, rather than social and economic problems existing pre-formed and independently 'out there' in society waiting for the state to act to try to solve them, the governmentality perspective suggests that the problems are themselves in part a product of state processes. A good example of this is provided by the phenomenon of 'problem regions'. From the point of view of theories of governmentality 'problem regions' do not have an existence independent of government efforts to identify and address them. This is not to say that some geography areas do not suffer from greater levels of poverty or unemployment than others. However, it is (usually) the state that defines what a region is and what features it has to have to constitute a problem worthy of state policy-making. As I have shown elsewhere, in the case of the recent development of new regional economic strategies in England, it was in part the texts of the strategies themselves that brought into being the regional economies that the strategies were supposed to develop (Painter 2002a).

These brief and disparate examples of feminist state theory, citizenship, mental illness and governmentality are drawn from work in political, sociocultural, historical and economic geography. They are by no means a representative survey of human geography, but they exemplify the growth of interest in the importance of the state in contemporary geographical scholarship. Most such accounts are not written explicitly or otherwise from either a strategic-relational or a world-systems perspective and not all of them transcend the conventional treatment of 'state' and 'society' as distinct interacting spheres. Nonetheless, they are testament to a growing commitment to taking the state seriously throughout human geography.

FURTHER READING

Poulantzas, Nicos (1978) *State, Power, Socialism.* London: New Left Books

Mitchell, Timothy (1991) 'The limits of the state: beyond statist approaches and their critics', *American Political Science Review* 85: 185–213

Taylor, Peter J (1996) 'Embedded statism and the social sciences: opening up to new spaces'. *Environment and Planning A* 28 1917–28
See also the responses to Taylor's paper in the same issue.

Peck, Jamie (2001) *Workfare States.* New York: Guilford

Jessop, Bob (2001) 'Bringing the state back in (yet again): reviews, revisions, rejections, and redirections. *International Review of Sociology* 11: 149–73

Jessop, Bob (2002) *The Future of the Capitalist State.* Cambridge: Polity Press

Brenner, Neil, Jessop, Bob, Jones, Martin and MacLeod, Gordon, (eds) (2003) *State/Space: A Reader.* Oxford: Blackwell

References

Agnew J (2002) *Making political geography*. London: Arnold.

Braun B (2000) Producing vertical territory: geology and governmentality in late Victorian Canada. *Ecumene* 7: 7–46.

Brenner N (1998) Between fixity and motion: accumulation, territorial organization and the historical geography of spatial scales. *Environment and Planning D: Society and Space* 16: 459–81.

Brenner N (2001) The limits to scale? Methodological reflections on scalar structuration. *Progress in Human Geography* 25: 591–614.

Burchell G, Gordon C, Miller P, eds (1991) *The Foucault effect: studies in governmentality*. Hemel Hempstead: Harvester Wheatsheaf.

Clark GL (1981) Democracy and the capitalist state: towards a critique of the Tiebout hypothesis. In *Political studies from spatial perspectives*, ed. AD Burnett, PJ Taylor, pp. 111–29. Chichester: John Wiley & Sons.

Clark GL, Dear MJ (1981) The state in capitalism and the capitalist state. In *Urbanization and urban planning in capitalist societies*, ed. MJ Dear, AJ Scott, pp. 45–62. London: Methuen.

Clark GL, Dear MJ (1984) *State apparatus: structures and languages of legitimacy*. Winchester, MA: Allen and Unwin.

Cox K (2002) 'Globalization,' the 'regulation approach,' and the politics of scale. In *Geographies of power: placing scale*, ed. A Herod, MW Wright, pp. 85–114. Oxford: Blackwell.

Cravey AJ (1998) Engendering the Latin American state. *Progress in Human Geography* 22: 523–42.

Dean M (1999) *Governmentality: power and rule in modern society*. London: Sage.

Dear MJ (1981a) The state: a research agenda. *Environment and Planning A* 13: 119–15.

Dear MJ (1981b) A theory of the local state. In *Political studies from spatial perspectives*, ed. AD Burnett, PJ Taylor, pp. 183–200. Chichester: John Wiley & Sons.

Dear MJ, Clark GL (1978) The state and geographic process: a critical review. *Environment and Planning A* 10: 173–83.

Evans P, Rueschemeyer D, Skocpol T, eds (1985) *Bringing the state back in*. Cambridge: Cambridge University Press.

Foucault M (1979) On governmentality. *Ideology and Consciousness*: 5–21.

Goldring L (2001) The gender and geography of citizenship in Mexico–US transnational spaces. *Identities: Global Studies in Culture and Power* 7: 501–37.

Gottmann J (1952) *La politique des états et leur géographie*. Paris: Colin.

Habermas J (1987) *The theory of communicative action. Vol 2: The critique of functionalist reason*. Cambridge: Polity.

Häkli J (2001) In the territory of knowledge: state-centred discourses and the construction of society. *Progress in Human Geography* 25: 403–22.

Hartshorne R (1939) *The nature of geography*. Lancaster, PA: Association of American Geographers.

Hartshorne R (1950) The functional approach in political geography. *Annals of the Association of American Geographers* 40: 95–130.

Harvey D (1976) The Marxian theory of the state. *Antipode: A Radical Journal of Geography* 8: 80–98.
Jessop B (1997) The regulation approach. *The Journal of Political Philosophy* 5: 287–326.
Jessop B (2000) The crisis of the national spatio-temporal fix and the tendential ecological dominance of globalizing capitalism. *International Journal of Urban and Regional Research* 24: 323–60.
Jessop B (2001) Bringing the state back in (yet again): reviews, revisions, rejections, and redirections. *International Review of Sociology* 11: 149–73.
Jessop B (2002) *The future of the capitalist state*. Cambridge: Polity.
Johnston RJ 1980. Political geography without politics. *Progress in Human Geography* 4: 439–46.
Johnston RJ (1982) *Geography and the state: an essay in political geography*. Basingstoke: Macmillan.
Jones MR (1997) Spatial selectivity of the state? The regulationist enigma and local struggles over economic governance. *Environment and Planning A* 29: 831–64.
Luke TW (1996) Governmentality and contragovernmentality: rethinking sovereignty and territoriality after the Cold War. *Political Geography* 15: 491–507.
MacKinnon D (2000) Managerialism, governmentality and the state: a neo-Foucauldian approach to local economic governance. *Political Geography* 19: 293–314.
MacLeod G, Goodwin M (1999) Reconstructing an urban and regional political economy: on the state, politics, scale, and explanation. *Political Geography* 18: 697–730.
Mann M (1984) The autonomous power of the state: its origins, mechanisms and results. *Archives Européennes de Sociologie* 25: 185–213.
Miliband R (1973) *The state in capitalist society*. New York, NY: Basic Books.
Mitchell T (1991) The limits of the state: beyond statist approaches and their critics. *American Political Science Review* 85: 77–96.
Moon G, Brown T (2000) Governmentality and the spatialized discourse of policy: the consolidation of the post–1989 NHS reforms. *Transactions of the Institute of British Geographers* 25: 65–76.
Murdoch J, Ward N (1997) Governmentality and territoriality – The statistical manufacture of Britain's 'national farm'. *Political Geography* 16: 307–24.
Painter J (1995) *Politics, geography and 'political geography': a critical perspective*. London: Arnold.
Painter J (2000) State and governance. In *A companion to economic geography*, ed. E Sheppard, T Barnes, pp. 359–76. Oxford: Blackwell.
Painter J (2002a) Governmentality and regional economic strategies. In *Habitus: a sense of place*, ed. J Hillier, E Rooksby, pp. 115–39. Aldershot: Ashgate.
Painter J (2002b) The rise of the workfare state. In *Geographies of global change: remapping the world*, ed. RJ Johnston, PJ Taylor, M Watts, pp. 158–73. Oxford: Blackwell.
Painter J (2003) Towards a post-disciplinary political geography. *Political Geography* 22: 637–9.
Peck J (2001a) Neoliberalizing states: thin policies/hard outcomes. *Progress in Human Geography* 25: 445–55.

Peck J (2001b) *Workfare states*. New York, NY: Guilford.
Peck J (2002) Political economics of scale: fast policy, interscalar relations, and neoliberal workfare. *Economic Geography* 78: 331–60.
Philo C (2004) *The space reserved for insanity: an historical geography of the 'mad-business' in England and Wales to the 1860s*. Lampeter: Edwin Mellen Press.
Poulantzas N (1973) *Political power and social classes*. London: New Left Books.
Poulantzas N (1978) *State, power, socialism*. London: New Left Books.
Raffestin C, Lopreno D, Pasteur Y (1995) *Géopolitique et histoire*. Lausanne: Editions Payot.
Ratzel F (1969) The laws of the spatial growth of states. In *The structure of political geography*, ed. RE Kasperson, JV Minghi, pp. 17–28. London: University of London Press.
Soja EW (1968) Communications and territorial integration in East Africa: an introduction to transaction flow analysis. *East Lakes Geographer* 4: 39–57.
Taylor PJ (1981) Political geography and the world-economy. In *Political studies from spatial perspectives: Anglo-American essays on political geography*, ed. AD Burnett, PJ Taylor, pp. 157–72. Chichester: John Wiley & Sons.
Taylor PJ (1996) Embedded statism and the social sciences: opening up to new spaces. *Environment and Planning A* 28: 1917–28.
Taylor PJ (2003) Radical political geographies. In *A companion to political geography*, ed. J Agnew, K Mitchell, G Toal, pp. 47–58. Oxford: Blackwell.
Taylor PJ, Flint C (2000) *Political geography: world-economy, nation-state and locality*. Harlow: Longman.
Wallerstein I (1974) *The modern world system. Capitalist agriculture and the origins of the European world-economy in the sixteenth century*. New York, NY: Academic Press.
Wallerstein I (1976) A world-system perspective on the social sciences. *British Journal of Sociology* 27: 345–54.
Wallerstein I (1979) *The capitalist world-economy*. Cambridge: Cambridge University Press.
Wallerstein I (1983) *Historical capitalism*. London: Verso.

4 Culture: Economy

Trevor Barnes

> Culture is not 'a decorative addendum to the "hard world" of production and things, the icing on the cake of the material world'.
>
> (Hall, 1988)

> Many of us have learned to want to cleave to an order. This is a modernist dream. In one way or another, we are attached to the idea that if our lives, our organizations, our social theories or our societies, were 'properly ordered' then all would be well. And we take it that such ordering is possible, at least some of the time. So when we encounter complexity we tend to treat it as a distraction. We treat it as a sign of the limits to order. Or we think of it as evidence of failure.
>
> (Law, 1994: 4–5).

What do Christmas, the Sex Pistols and a plank of wood have in common? While this sounds like the beginning of a bad joke, I want to use the question as an entrance point to one of the more intractable issues in the social sciences, the relation between economy and culture, and which, with its recent 'cultural turn', spills into economic geography (Crang, 1997; Thrift, 2000a; Barnes, 2001). In academic and popular representations, economy and culture are often sharply separated, put into quite different conceptual boxes. Karl Marx, for example, puts economy in a box called base or infrastructure, and culture in a very different box called superstructure. Or Talcott Parsons puts economy in a box called adaptation, and culture in another box called latency. This same separation is also repeated in popular media: newspapers separate out their business and review sections; there are magazines focusing on the economy (for example, *The Economist Magazine*) and magazines focusing on culture (such as *People Magazine*); and there are TV programmes that discuss only finance (for example, 'Wall Street Review') and other shows that are only concerned with the arts (such as 'Sister Wendy's One Thousand Masterpieces').

Indeed, in North America there are whole channels devoted exclusively to one or the other topic, 24 hours a day, seven days a week.

In contrast, I argue in this chapter that rather than being clear and distinct, the relation between economy and culture is muddy and indefinite. Following the argument of this book, I suggest that economy and culture are not a neatly divided binary but a hybrid. This is also the answer to my opening question. What Christmas, the Sex Pistols and a plank of wood have in common is that they are hybrids of culture and economy. That is, they take on attributes of both such that it is difficult to know where one begins and the other ends.

That Christmas is a culture-economy hybrid is clear (and discussed systematically by Thrift and Olds, 1996). Marking the birth of one of the world's greatest cultural religious figures, Christmas in high-income Western countries is also fundamentally about money: it is about shopping until you drop, crowded retail malls, sales and bargains, gifting, re-gifting and de-gifting. It is about spending time with family and friends, but it is also about overdrawn bank accounts and usurious interest on unpaid credit card balances. It is about eating special foods – in my Anglo-cultural tradition such delicacies as mince tarts, sausage rolls and Christmas pudding – and drinking special drinks – egg nogg, sherry and warm punch – and carrying out culturally specific rituals – trimming the tree, lighting the Christmas pud, putting up interior and exterior house decorations – but it is also about capitalist producers drawing together an immense amount of economic resources, and undertaking planning, sometimes years in advance, for the production, delivery and selling of the kind of commodities required to make these rituals possible, such as turkeys, or Christmas trees, or Christmas cards and wrapping paper. It is about glistening eyes and radiant smiles at infant school nativity plays or at elementary school Christmas concerts, or hissing at villains and cheering at heroes in pantomimes, or seeing Christmas specials on the TV that you have seen ever since you were a child – Alistair Sim as Scrooge terrorizing the Cratchett family, Mr Bean getting his head stuck in the rear-end of a turkey, James Stewart realizing 'it's a wonderful life' – but it is also about unrelenting TV adverts, fliers in newspapers, handbills in letter boxes, under car windshield wipers, and thrust into your fingers on the street, all trying to convince you to open your purse and wallet to buy that perfect commodity for your loved ones. Is Christmas a cultural celebration? Or is Christmas a once-a-year economic bonanza for capitalism? It is both. It is a hybrid.

The same goes for the Sex Pistols, the English punk-rock band that burst on to the cultural scene in 1977 with their singles 'Anarchy in the UK' and 'God Save the Queen', and whose LP 'Never Mind the Bollocks. Here's the Sex Pistols' initially was sold under plain brown wrapping so as not to offend innocent bystanders. Apart from the fact that they produced a cultural product, music (or at least *they* thought so – Sid Vicious said, 'You just pick a chord, go twang, and you have music'), the Pistols influenced, and came to define, a wider youth sub-culture, punk. It was a culture defined by: particular forms of dress and deportment – ripped jeans and t-shirts, green- and red-dyed hair, Doc Marten boots, and the ubiquitous use of safety pins for tethering things that should never be tethered; a specialized language and vocabulary marked by 'insults and rejection ... [but] yet recognisable as everyday speech' (Savage, 1993: 206); and a broader philosophy and politics of life rooted not in 'the negative or cynical rejection of belief but the positive courage to live without it' (Savage, 1993: 195). If the hallmarks of culture are innovation, new forms of language, and changed values and ways of life, then the Sex Pistols were real culture. The Pistols from their very creation, however, were always also about making money, of selling product, of being part of 'the great rock 'n' roll swindle' (Mitchell, 2000: 68). They were a manufactured band directed towards making profit for their sponsors, primarily their manager Malcolm McLaren. As a consequence, it was entirely in fiscal character that the Pistols recorded with then-on-the-run great train robber Ronnie Biggs, that Johnny Rotten engaged in an eight-year legal suit with McLaren to recover unpaid royalties (Lydon, 1994: chs 19–20), and that McLaren coined the slogan 'cash from chaos', and also insisted that 'Sex' be in the name of the group because it advertised his shop of the same name located on Kings Road, Chelsea. Were the Sex Pistols an explosive and original cultural phenomenon, 'a distinct break in the pop milieu' (Marcus, 1989: 2–3), like Picasso's cubism, or Schoenberg's atonalism, or Joyce's stream of consciousness, or were they only about 'filthy lucre'? They were both. They were a hybrid.

Making the hybrid argument for a piece of lumber is more of a stretch. Like 'death and furniture', to use Edwards, Ashmore, and Potter's (1994) examples, there is a seemingly brute finality about a plank of wood that puts it outside of the cultural pale. A plank is a plank is a plank. But it is not. Culture infuses every stage of plank-making and selling, which I will illustrate by using examples from one of the great plank-making regions of the world, British Columbia (BC). It begins with broad cultural attitudes to nature, and in the case of BC an attitude towards the temperate rainforest that covers the province's southern coastal region. Are we stewards of this forest or masters of it? Since European settlement in the mid-nineteenth century mastering has been the order of the day. Legally formalized through the recommendations of a series of BC Royal Commissions on forestry beginning in 1945, mastering nature in the province has produced gargantuan clear-cuts, a policy of cut and run and the proliferation of single-industry towns, one task of which is to produce two-by-fours. Culture further extends into the very process of work. While recently those work practices have been transformed following a move from Fordist to post-Fordist methods (Barnes et al., 2001), they remain highly masculinized – 'men's work' – turning on the use of large machines and male brawn (about 90 per cent of jobs in the Canadian wood products are held by men: Randall and Ironside, 1996). Furthermore, the masculinist culture inside the sawmills infects the culture outside, producing among other things patriarchal relations (Egan and Klausen, 1998), a particular culture of education, especially among boys who discount its benefits because of the presumed inevitability of job availability at the mill (Behrisch et al., 2003), and a community culture of fatalism, feelings of impermanence and isolation, and characterized by a lack of initiative. Finally, culture stretches into the kinds of planks that are made. In particular, from the 1980s onwards, among especially coastal BC lumber producers, there was a switch from selling to the US and Canadian housing construction industry to exporting, particularly to Japan, South East Asia and China. But those latter markets do not want traditional two-by-fours, but lumber of quite different dimensions to construct their own culturally distinct housing and commercial buildings (Hayter and Edgington, 1987). In part, the shift from Fordist to post-Fordist methods was to enable production of these new dimensions. So, is a plank only a plank? In my interpretation, it is not. While a plank might be physically unblemished, it is knotted by both economic and cultural markings; it is a hybrid.

The argument in this chapter is that the economy–culture hybrids illustrated above are pervasive in economic geography. That we do not typically recognize them is because of an equally pervasive mindset of binary thinking found in the discipline that divides phenomena into either the economic, which is a legitimate domain of study, or the cultural, which is not, and is passed on to others to investigate. With the 'cultural turn' in economic geography such a mindset is not viable any more, if ever it was, and we need to engage in different practices. To determine what kind of practices, I divide the chapter into three parts. First, I discuss the longstanding debate about the relationship between economy and culture, and economic geography's relationship to it. What emerges is a need to theorize outside the binary of economy and culture. Secondly, I will argue that such a theorization is achievable by using the idea of a hybrid, and here I draw principally on the work of Bruno Latour and other proponents of actor-network theory. Finally, I discuss two attempts by economic geographers – respectively J.K. Gibson-Graham and Nigel Thrift – to work through the idea of an economy–culture hybrid, and which can be used as the basis for a potentially reconfigured economic geography.

Economy, culture and economic geography

Economy and culture are typically interpreted as a binary. For Derrida such binaries are the very vehicle by which meaning is created in the West, forming the basis of what he calls logocentrism (Gibson-Graham, 2000). Logocentrism – meaning literally reason-centred – is the strategy of producing meaning by dividing language into opposite pairs of terms. The first term of a given binary is defined positively, and the second term is defined as its antonym. Formally, it is represented by the scheme A/not-A. In the case of economy and culture, the first term, economy, is positive, and the second term, culture, is everything that the economy is not. As a result, while the economy is hard, culture is soft; while the economy is about facts, culture is about values; while the economy is strong, masterful and masculine, culture is weak, submissive and feminine. And on it goes.

It further follows given the binary nature of this scheme that economy and culture are doomed never to be on the same page; that things economic are separated from things cultural because they are their opposites. Certainly, such a strategy characterized much of social science, which in the past assiduously divided economy and culture into sealed spheres of enquiry.

In particular, the study of the economy is given over to the discipline of economics. In its orthodox form, known as neoclassicism, economics removes any trace of culture from its study by making two related methodological moves. The first is by pursuing methodological individualism, an approach that explains social events by reducing them to the beliefs and actions of the individual rational actors involved. More specifically, in the neoclassical model, the economy is constituted by a set of rational individuals. Imbued with an exogenous map of preferences, those individuals are guided in their consumption and production choices by both their innate rationality and the invisible hand of market prices that ensures eventual collective optimization and harmony. Following methodological individualism, that collective never exists, *sui generis*, on its own, but is only ever the sum of its individual parts; that is, it is always reducible to the beliefs and actions of constitutive individual rational agents. Consequently, methodological individualism portrays culture as chimera, as something that appears real, but is not. As Margaret Thatcher might have said, 'There is no such thing as culture, only individuals.' The real are rational economic agents, the spurious are communal entities like culture.

The second methodological move evacuating culture is neoclassicism's deployment of formal modes of reasoning and analysis. Joan Robinson once said, 'You can't put culture into an equation.' If so, neoclassical economics has no hope of dealing with culture because as a discipline it defines itself by the use of mathematical models. For example, the influential American economist Paul Krugman says, 'to be taken seriously an idea has to be *something you can model*' (Krugman, 1995: 5; original emphasis). And while admitting that 'people who do not write formal models may have rich insights', he also says, 'strangely, though, I can't think of any' (Krugman, 1995: 88). The further implication is that if non-economists, such as economic geographers, are to study the economy then to have credibility they must also adopt the standards of economics and undertake formal modelling. A failure to do so results in economic geographers not studying the economy, but engaging in forms of pseudo enquiry. This is Krugman's opinion of current research by economic geographers who eschew formal modelling. He lambasts those, for example, who make reference to '"post-Fordism" ... [a term coined by] the Derrida-influenced regulationist school – deconstructionist geography!' (Krugman, 1995: 85). While it is doubtful that Derrida is much of a regulation theorist (Eagleton, 1995), the important point for Krugman is that by shunning formal models economic geographers engage in absurd forms of study like 'deconstructionist geography!'. It is the economist's way or no way. And the economist's way produces a distilled world of pure economy, unblemished by culture.

Neoclassical economics is extreme in its aversion to culture, leaving its study to soft, unscientific, or non-rigorous social sciences and which include, at least in Krugman's interpretation, economic geography. It is true that during one period in the history of economic geography, some but certainly not all economic geographers threw their lot in with the economists, and by deploying neoclassical mathematical models, and adopting methodological individualism, pursued an approach that displaced culture. The most systematic application of neoclassical economic principles to economic geography occurred roughly in the 20-year period 1955–75, and was associated with the discipline's 'quantitative revolution' (Barnes, 2001). Best represented by the theoretical tradition of locational analysis (Haggett, 1965), culture was often difficult to discern in the consequently flattened, formalized economic geographical geometries of concentric rings of agricultural production, industrial locational triangles, and hexagonal market nets of central place services (Barnes, 2003).

Some of the work within locational analysis represented a direct translation of the methodological individualist assumption of the rational agent to the geographical case. For example, in the writings on spatial interaction, which was often formulated mathematically in terms of the Newtonian gravity model equation, it was assumed that individuals who spatially interacted exhibited 'uniformly rational behavior' (Sheppard, 1978: 388). In particular, individual decisions to interact spatially were determined by their utility function, which they attempted to maximize. But the form of that function, and which represents a person's preferences for travel, and for that matter everything else, is given exogenously; that is, an individual's preferences are represented as lying outside formal analysis, neither requiring nor obtaining scrutiny and explanation. But it is those preferences that contain everything interesting about cultural norms, values and rules. It is in this sense that economic geography in the tradition of locational analysis only begins after culture is set aside.

Locational analysis, and the influence of neoclassicism, began waning in economic geography from the mid-1970s when a new, political economy approach increasingly took hold (albeit not without resistance; Johnston, 1991). At least initially, however, the same binary impulse remained, one that cleaved economy and culture, and asserted the dominance of the former over the latter. To see how and why, it is best to return to Marx's most succinct, and most well-known theoretical statement about economy and culture, and found in his Preface to *A Contribution to the Critique of Political Economy* (1859). There Marx writes, 'the mode of production of material life conditions the social, political and intellectual life process in general. It is not the consciousness of men that determines their being, but, on the contrary, their social being that determines their consciousness' (Marx, 1904: Preface). As Marx makes clear in other parts of the Preface, the social being that determines consciousness is rooted in 'relations of production which correspond to a definite stage of development in the material productive forces' (Marx, 1904: Preface). It is therefore economic relations, or what Marx called the 'base' or 'infrastructure,' that is the prime mover. In contrast, social being or consciousness, which correspond to the 'infrastructure' and comprise cultural elements such as religious and legal institutions, are the consequence and of secondary importance.

Marx's position is often interpreted as economic determinism: culture as a set of 'social, political and intellectual processes in general' is irrevocably determined by the economy, 'the mode of production'. Culture is thereby reduced to an epiphenomenon, performing the functional role of an ideological smokescreen for an oppressive capitalist class bent on immiserizing the proletariat. In this interpretation, while there are differences between Marx's and neoclassicism's view of economy and culture – for Marx culture performs a functional role in keeping capitalism going, but for neoclassicism it plays no role – there are also strong commonalities: both prioritize and essentialize the economy, conceiving it respectively as either the embodiment of material productive forces or of individual rational choice. Culture, in contrast, is sloughed off; it is not real, not essential and not a priority.

Certainly, when economic geographers began systematically drawing upon Marx's work they reproduced the economic determinist interpretation of classical Marxism, where culture was an after-thought if it was a thought at all. Such a position is perhaps best found in the work of David Harvey, who through the 1970s and 1980s developed a systematic and often brilliant geographical interpretation of Marx through a detailed exegesis of his writings (summarized in Harvey, 1982). The details are complicated, but the gist is that propelling geographical change are economic imperatives of accumulation turning on the maximization of surplus value within the sphere of production. Culture does not get much of a mention by Harvey until the late 1980s, when it is theorized in *The Condition of Postmodernity* (Harvey, 1989), and subtitled significantly, *An Enquiry into the Origins of Cultural Change*. But even here, Harvey resorts to a version of Marx's original base–superstructure model. Modernist culture is predicated upon a Fordist economy, and postmodernist culture is predicated upon a post-Fordist one. Furthermore, both types of economies derive from the basal logic of all capitalist accumulation resting on an incessant need to reduce turnover time in order to realize surplus value. Marx's original binary thus remains firmly in place. Ten years later or so, still not much has changed. Harvey in his introduction to his latest book, *Spaces of Hope* (2000), continues to assert the dualism of culture and economy. On the one hand, there is 'cultural analysis' which is 'fun', and on the other, 'political economy' which is about 'the dour world and crushing realities of capitalist exploitation)' (Harvey, 2000: 5). For Harvey (2000: 7), of course, it is those 'crushing realities' that demand our attention, and his list of the most important – 'fetishism of the market', 'the savage history of downsizing', 'technological change', 'weakened organized labour' and an 'industrial reserve army' – make it clear that it is not going to be fun. We need to be prepared for serious work. No cultural analysis here.

I realize this is caricaturing Harvey, and that his prose and theorizing is more nimble and nuanced than I have suggested. But he is now almost alone among Marxist geographers in his resistance to 'cultural analysis'. For example, the former dyed-in-the-wool, classical Marxist Dick Peet (2000: 1215) now searches for 'the cultural source of economies', urging the use of 'cultural terms such as symbol, imaginary, and rationality ... to understand crucial economic processes' (p. 1213). As Peet (2000: 1231) writes, 'In a phrase I never thought I would say, political economy should become cultural economy.' Or Neil Smith, a student of Harvey's, who in his earlier days trumpeted 'the universalization of value in the form of abstract labour' (Smith, 1984: 82) now says, '"Back-to-class" in any narrow sense is its own self-defeating cul-de-sac' (Smith, 2000: 1028), and it is necessary 'to find a way of integrating class into the issues of identity and cultural politics' (Smith, 2000: 1011).

Helping produce such changes of heart among former hard-line Marxist economic geographers has been an increasingly large body of work from the left that rethinks the relationship between economy and culture, and in doing so moves away from Marx's binary base–superstructure relation. Again, the details are complex. Its general impetus is to show that the cultural, for example, as gender, and/or race, and/or religion, partially constitute(s) the economic and vice versa, and as a result there is no neat dividing line between one sphere and the other. Although there are historical precedents for this cultural turn (for example, in Gramsci's notion of 'hegemony' articulated in the 1920s), much of the impulse derives from cultural studies, which comes in at least two versions. The older British type, and particularly associated with Raymond Williams ('structure of feeling'), Richard Hoggart ('the felt quality of life') and later Stuart Hall ('Marxism without guarantees'), remains committed to a socialist politics intent on remedying the inequities of capitalism. Its novelty is in trying to hang on to class analysis and the economy, while at the same time recognizing and merging cultural values and practices, ways of life, and emotional and political commitments that lie outside. Hence, for example, Williams' phrase the 'structure of feeling' that connotes the 'doubleness of culture ... [as both] material reality and lived experience' (Eagleton, 2000: 36). The North American version of cultural studies is less interested in class politics. Embracing tenets of post-structuralism, especially as expressed by Derrida and Foucault (Gibson-Graham, 2000), culture and economy can become in the North American version simply the 'free play of texts, representations and discourses' (Bradley and Fenton, 1999: 114).

In economic geography, the move towards culture has been a long time coming. Of course, one might argue that the two were together right from the beginning. Certainly, the line between the economy and culture is smudged, if it is there at all, in early economic geographical texts such as in Chisholm (1889) or in J. Russell Smith (1913). That said, it is not until Doreen Massey's (1984) *Spatial Divisions of Labour* that there is an explicit attempt to *theorize* the relation between culture and economy. Writing from a Marxist perspective, Massey in effect criticizes, and provides an alternative to Harvey's theory of accumulation. Using a geological metaphor, she conceives of culture and economy in particular places as involved in a recursive, two-way relationship, and in so doing, she clearly moves away from Marx's classical base–superstructure rendering. There is not space to provide a detailed account of her analysis. The important point is that *Spatial Divisions* put culture on the economic geographer's theoretical agenda in a way that it was never there before, and paved the way for the 'cultural turn' now found in the discipline. That 'cultural turn' has taken a variety of forms, embracing both the traditions of British and North American cultural studies, but it has also drawn upon other bodies of work as well, ranging from critical realism (Sayer, 1997), to Karl Polyani's idea of embeddedness (Mitchell, 1995), to Thorstein Veblen's institutionalism (Martin, 1994).

All of this is to say, that in the past few decades there has been a sustained effort, at least from some factions on the left, as well as the left in economic geography, to shift their thinking from the binary of culture and economy to something more complex; one that tries to overlap and dissolve the boundaries of the two categories. To use the vocabulary of this book, there has been a move to conceive culture and economy as a hybrid.

Hybrids and translation

Deriving from botany and zoology, and more recently genetics, a hybrid is the result of combining two different species or different genetic strains of the same species. The consequence is a new object that, while sharing characteristics of its 'parents', also possesses unique traits.

That hybridity as an idea has travelled from the life sciences to the social sciences is because of dissatisfaction with the conceptual binaries pervading the latter. Hybrids provide the possibility for thinking outside traditional dualisms. Additionally, such possibilities have political resonance, for example, as is the case with the use of hybridity in post-colonial literature. The identity of the colonial migrant living within the metropole is hybrid, lying betwixt and between colonial and native subject positions. As such, it challenges the very binary of home and colony, the same and the Other (Bhabha, 1994), and the relations of power that inhere between them. Or again, Donna Haraway (1991) uses the notion of a cyborg – a creature that is half human and half machine – as an exemplar for the kind of politics that she wants to prosecute, and which cuts across hitherto oppressive binaries like man/woman, straight/gay, white/black and able-bodied/disabled. The hybrid cyborg interrupts and thwarts categories typically used to assert domination, power and control. As I will discuss in the next section, some economic geographers also believe that opening up a hybrid space between economy and culture provides a location for political potentialities.

Another possibility afforded hybrids is making connections between types and objects of enquiry that because of binary thinking have been hitherto kept separate, for example, nature and society, science and humanities, or in our case, economy and culture. In particular, the notion of a hybrid or hybrid *collectif* has been used extensively within actor-network theory (ANT) to subvert the dualism of nature and society (Latour, 1993; Callon and Law, 1995). By reviewing that work, I will try to provide a vocabulary and set of concepts that I can use in the last section to discuss a hybridized economy and culture within a reconceived economic geography.

ANT emerges out of science studies, and is especially associated with the work of Bruno Latour (1993, 1999) and John Law (1994, 1999). One of its most insistent claims is that 'entities take their form and acquire their attributes as a result of their relations with other entities' (Law, 1999: 3). Those entities can literally be anything – Petri dishes, microscopic bacteria, spectrometers, papers published in *Nature*, people in white coats. There are no limits. The important point is that the meaning of an entity is a relational effect of its association with other entities. As Callon and Law (1995: 485) write: 'it is the relations ... that are important. Relations which perform.' In the language of ANT, individual entities are called actants, and the sum of the relations established among them is termed a network. A network is a tight-knit assemblage of heterogeneous actants that are persuaded at least temporarily to stay in place, and work with one another to produce order (Whatmore, 1999: 28; Murdoch, 1997).

Order is only as durable as the network itself, however, and reaches only as far as the network extends. There is nothing outside. Order is not universal and fixed from on high, given by transcendental categories like Nature or Society, or Economy and Culture. Rather, it is contingent upon the workings of particular networks, their stability and geographical reach. As Whatmore (1999: 31–2) writes, order is ' always in the making, not ... a priori'. That we sometimes think it is a priori, inherent in transcendental categories, is because of a process of purification (Latour, 1993); that is, the process of fastidiously dividing the world into unblemished categories – rocks into Nature, working class into Society, money into Economy, opera into Culture – making the world neat and tidy, as if it is one giant filing cabinet. But for Latour such categories are not beginning points for enquiry, but end points. As he writes, Society and Nature are not 'causes of our knowledge ... but a consequence'. So, scientists do not start with Nature, but produce it only at the end of their research by a process of purification; that is, by systematically excluding from their accounts parts of their work that do not fit into the Nature box. But in reality scientists continually bring together all manner of heterogeneous entities – Petri dishes, microscopic bacteria, and so on – connecting them, forming networks, or what amounts to the same thing, working with, and producing hybrids. But those hybrids are not recognized. They are continually purified into sub-components, separated into various binaries. Latour (1993: 2) provides an example:

> The smallest AIDS virus takes you from sex to the unconscious, then to Africa, tissue cultures, DNA and San Francisco, but the analysts, thinkers, journalists and decision makers will slice the delicate network traced by the virus for you into tidy compartments where you will find only science, only economy, only social phenomena, only local news, only sentiment, only sex.

For Latour, however, if we want to understand how science works, we must trace through the multitudinous networks of which it is composed, recognizing their hybridity. The real action does not take place at the rarefied poles of Nature and Society, but in 'the middle kingdom' (Latour, 1993: 47); that is, within the 'imbroglio' or 'Gordian knot' of hybrids (Latour, 1993: 3). Callon and Law (1995: 485) call that imbroglio a 'hybrid *collectif*', and define it as 'an emergent effect created by the interaction of the heterogeneous parts that make it up'. As with Latour, they are keen to stress that the hybridity of the *collectif* often goes unrecognized, and, important for this chapter, it is a criticism that they apply as much to social scientists as to natural scientists. Social scientists are as blinkered by binaries as natural scientists. The solution is to follow the interaction of heterogeneous actants wherever they lead, and the hybrids that are consequently produced.

Useful for understanding how hybridity is produced from these heterogeneous actants is the notion of 'translation'. Michel Callon (1980: 211) writes that 'translation involves creating convergences and homologies by relating things that were previously different'. Translation is thus the very basis of hybridity. Latour (1999: 88) elaborates:

> The question of translation consists of combining two hitherto interests ... to form a single goal ... Even if the balance is equal, neither of the parties ... will be able to arrive at exactly [their] original goal. There is a drift, a slippage, a displacement, which, depending on the case, may be tiny or infinitely large.

Translation, then, involves bringing together entities that are sometime radically different, and convincing them that they have an interest in connecting and relating. Through a process of 'drift', 'slippage' and 'displacement', that connection creates something new, a hybrid. Callon's (1986) original example of translation involved marine scientists, fishers, scallops and techniques of scallop propagation in St Brieuc Bay in Normandy, France. The context is a sudden decline in scallops in the Bay for harvesting. The task of the scientist is to understand the process, and which, as Callon shows, is achieved through creating a hybrid *collectif* that brings together through translation the four actants. Scientists must be persuaded that St Brieuc is an interesting problem to study, fishers must be persuaded to curb their harvesting while scientists work, scallops must be persuaded to join the experiments of the scientists, and techniques of scallop propagation must be persuaded to operate in conditions quite different from those in which they were initially designed. There are no guarantees here, and to enable the hybrid *collectif* to emerge and attain stability requires the expenditure of a lot of work and resources. It is always in process, always an achievement.

In sum, the science studies literature on hybrids provides a vocabulary and set of concepts to represent the relationship between culture and economy. It says that terms like Economy and Culture are the consequence of a prior process of purification in which tainted hybrid entities are either removed, or cleansed. The world we study, though, is not like that. It is full of messy distinctions and objects that smudge boundaries and cross borders. It is a hybrid world. Michael Mann (1986: 1) writing in a different context says, 'the world is messier than our theories of it'. ANT provides a means of coping with the messiness. It says that we need to attend to the networks of relations that crosscut, interleave and fold across culture and economy, and that form hybrid collectives. More particularly, the task is to trace specific translations of objects, people and ideas, as they come together from a variety of origins, and sometimes split apart. It is by undertaking this tracing that we enter the 'middle kingdom', and see the world before it is torn in two, in this case purified by the binary of Culture and Economy.

In addition, there is a political imperative. I have not discussed issues of power, but they are there, especially in Law's (1994, 2001) work, and certainly in Haraway's (1994, 1997), who is sympathetic to ANT (her notions of a 'cat's cradle' and cyborg compliment the hybrid *collectif*). The political problem with applying purified categories, especially in a binary form, is that they can produce dogma and intolerance, and sometimes much worse (Bauman, 1989, brilliantly discusses the 'much worse'). It is for this reason that Haraway (1991: 81) says, 'I would rather be a cyborg than a Goddess.' Recognizing and staying with hybrids, with complexity, if not creating a kinder, gentler world, might create a more politically tolerant, modest one, less skewered by immaculate binaries. As Whatmore (1999: 35) writes, 'hybrid geographies cannot be other than plural and partial', which is surely better than monolithic and absolute.

Toward a hybrid economic geography: two examples

As I suggested, since Massey's (1984) work there has been a shift towards recognizing the cultural within economic geography, and since the mid-1990s that move has sped up producing the 'cultural turn', thus pushing the discipline farther away from the economism of either the neoclassical or Marxist type. Thrift (2000a: 692) even says that 'it is possible to argue that economic geographers have become some of the leading exponents of cultural geography'. However, there are already critics, including those who are otherwise sympathetic to the inclusion of the cultural within economic geography. For example, Martin and Sunley (2001) and Plummer and Sheppard (2001) in reacting to the cultural economic geographical manifesto put forward by Amin and Thrift (2000) see that project weighed down by the same kinds of problems found in economistic accounts except here it is a culturalist dogma that is espoused: actors are cultural rather than economic dupes. Further, both sets of critics are suspicious about both the lack of theoretical and empirical rigour in cultural approaches – Martin and Sunley (2001: 152) talk about its 'vague theory and thin empirics' – and the limited, small-scale nature of the studies produced that are unrepresentative of larger-scale phenomena and geographies.

Certainly, the arguments of Martin and Sunley and Plummer and Sheppard are important warnings. The task of a hybrid economic geography should not be simply to reverse the binary and make culture primary, and the economy secondary. Rather, it should be to dissolve the binary altogether (and which I would argue Thrift attempts at least in his substantive work, and discussed below). In addition, a hybrid economic geography should recognize and utilize a variety of methodological strategies, quantitative and qualitative, statistical and story-based, and not be methodologically monist. Finally, it should not confine particular kinds of phenomena to particular scales of analysis. The cultural is not inherently local, and the economic is not inherently global. Some hybrids of culture and economy, like LETS schemes, operate at a local scale, while other hybrids of culture and economy, like transnational corporations, operate at a global scale. Rather than assuming economy and culture are naturally found at particular geographical scales, the important task is to evaluate the strength and nature of the linkages of the particular hybrid in question to ascertain its ability to extend spatially its reach.

More broadly, the debate between Amin and Thrift, and Martin, Sunley, Plummer and Sheppard (and there were others, see the special issue of *Antipode*, volume 33, number 2, 2001) suggests that the task of bringing together culture and economy within economic geography is still in process, still in discussion. There are no definitive approaches or answers. There are already, however, some interesting substantive works that attempt to realize the promise of a hybrid economic geography that takes culture seriously, and to finish let me provide two examples.

The first, and exemplifying the political possibilities of a hybrid economic geography, is J.K. Gibson-Graham's (1996, 2000) work. Rejecting the classical Marxist base–superstructure position and associated economism they had earlier deployed, from the early 1990s they began making use of Althusser's concept of overdetermination, which enabled them to explore the political possibilities of conceiving the economy as an open, porous entity, influenced as much by culture as it influences culture itself (Gibson-Graham, 1996: Preface). By overdetermination, they mean the idea that everything determines everything else. The result is that it is impossible to uphold traditional binaries like culture and economy. The two are continually in interaction and exchange, displacing, drifting and swerving, creating new hybrid entities. Gibson-Graham (2000) give the example of the factory. As a term, it is part of a classic binary positively associated with economy, and all the other words that go with it like production, man and reason (Gibson-Graham, 2000: 98). As a master signifier, it defines itself by what it is not – household, reproduction, woman, emotion. Gibson-Graham argues, however, it is not this simple. The two sides of that binary are always leaking into one another, disrupting them. For example, it is possible to 'reverse the flow of cultural valuation by pointing out how many hours are spent in unpaid domestic labor in the household and how this contributes to Gross Domestic Product, [and which] if measured, would outweigh the labor performed in factory-based production' (Gibson-Graham, 2000: 98). Through this reversal, the binary begins to crumble.

They apply this same anti-binary, overdetermination logic to understand capitalism itself, and it is here that they work out the political possibilities of hybridization. Capitalism as a term gains meaning from being on the same positive side of the binary as factory, and associated with the kindred words already listed. The problem with this binary politically is that it makes capitalism seem invincible, masterful in every sense. People accept the binary, and act accordingly, believing they cannot go against it. The purified category of Capitalism is too strong to resist. But for Gibson-Graham (as for Latour), the political task is to tarnish those purities, and in doing so provide the possibilities for resistance and change. This needs elaborating. Gibson-Graham argues that once capitalism becomes Capitalism through the binary, the only means of transformation is seemingly an apocalyptic one, a revolution on the same scale and magnitude of Capitalism itself. But the effort and resolve necessary for such a revolution is so large as to preclude it. If, however, we begin to reconceive the binary through an overdetermination logic, and exemplified by the factory example, we begin to see that Capitalism is plural and partial – a hybrid – and as a result strategies for change appear far less daunting, and more likely to occur. In particular, Gibson-Graham provide two particular strategies for recognizing the hybridity of capitalism, thereby opening spaces for political action.

The first is empirical. It is to investigate alternative forms of capitalism. By their very nature as alternatives, they are necessarily open to features on the other side of the binary, for example, reproduction, household, family, emotion and culture. Here, Gibson-Graham has studied the 'alternative' economic regions of the Mandragon region in Spain, the Latrobe valley in Australia and the Katahdin region in Maine (for further details on their project see www.arts.monash.edu.au/projects/cep/knowledges). The second is conceptual. It is to replace traditional metaphors stemming from the old binary and used to describe capitalism – masculinity, heterosexuality, domination, closure – and which make it so monolithic and absolute, with another set turning on femininity, homosexuality, resistance and openness (Gibson-Graham 1996: chs 5, 6). Once capitalism is described by these new metaphors, political action and the promise of alternatives become imaginable in ways inconceivable before.

The second example comes from Nigel Thrift's work and his attempt to deal with the hybridity of particular quaternary service activities within capitalism. Thrift's work is part of a wider body of writing by both economic geographers and others that recognizes a sea change in the operation of capitalism over the past two decades, transforming the nature of goods produced and sold, the behaviour and choices of consumers, and the very internal work-a-day operations of firms. Here Scott Lash and John Urry's (1994) book on 'economies of signs', Ulrich Beck's (1992) work on 'reflexive modernization', and in economic geography, Nigel Thrift's (1999) own writings on 'soft capitalism' are signal contributions. Such works argue that the economy operates as a discursive construction blending economy and culture. As Thrift (1999: 136) says, 'capitalism seems to be undergoing its own cultural turn as increasingly ... business is about the creation, fostering, and distribution of knowledge'.

It is the 'knowledge' industries on which Thrift has focused, and in particular those in the financial sector and high-level management consultancy firms (Leyshon and Thrift, 1996; Thrift, 2000b). In both cases, he is keen to illustrate their hybrid nature, and he does so by drawing explicitly upon the vocabulary and concepts of ANT. The contrast he makes is between an earlier form of business organization and management resting on structures, hierarchies, systems of inputs and outputs, and control devices, and a newer form couched in the vocabulary of networks, post-bureaucracy, teams, virtual management and post-structuralist organization (Thrift, 1999). The earlier form represented an application of binary thinking, or pure categories, of seeing the firm as a large flow chart with its different operations neatly organized into pre-assigned separated boxes. This was the era of the 'rational company man' (Thrift, 1999: 154). But in the post-Bretton Woods, post-Cold War, post-Fordist period, with flattened corporate structures, pervasive use of IT and ever-compressed space–time horizons, binary thinking no longer works. There is a need to be looser, quicker, more flexible, more creative and more adaptable. And to be this, argues Thrift, requires above all openness to a heterogeneous range of institutions, techniques, technologies, ideas and people.

The task of 'corporate social persona of the 1990s' (Thrift, 1999: 154) is to bring these different elements together, to make them perform. That is, it is to make them work as a hybrid *collectif*. As a result, and for reasons already given, the distinction between culture and economy loses its force.

Concluding remarks

As John Law (1994: 4–5) says in the epigraph to this chapter, we become nervous in the face of complexity, treating it at best as a 'distraction', and at worst as a sign of 'failure', both ours and the wider system. One means by which we impose order is by deploying binaries of various kinds. They enable us to sleep at night, to have the modernist dream. My argument in this chapter, however, is that the binaries are the dream. The world is not cleaved so neatly. Shakespeare has Hamlet declare to Horatio, 'There are more things in heaven and earth, ... than are dreamt of in ... philosophy'. Shakespeare (and Hamlet) were probably not thinking of a critique of binaries when they said that, but they might have been. The contention of this chapter is that there are more things in heaven and earth, including within economic geography, that are hybrids than binaries: like Christmas, like punk rock, like a plank of wood. And they demand a vocabulary sensitive to their hybridity.

Such a vocabulary has not existed in economic geography for much of its recent history, which was dominated by the economism of either neoclassicism or classical Marxism. This is changing with the 'cultural turn'. One component of that 'turn', I suggested, might be a vocabulary drawn from ANT, and one which is sensitive to hybridity. That said, it is not the only sensitive vocabulary available, as is evident from other contributions to this volume. It is flexible, though, which I tried to illustrate by casting the work of Gibson-Graham and Thrift within it as they prosecute a hybrid economic geography.

Haraway (1991: 129) argues that for political and intellectual reasons we must be vigilant in search of 'geometries, paradigms and logics [that] break out of binaries'. But even if we find them, such as ANT, there are always pressures to capitulate and return to a binary world. Jonathan Murdoch (1997: 732) writes: 'Spanning the divides, overcoming the dualisms, will not simply be a matter of adding terms such as "hybrid" or "cyborg" into our existing modes of thought, but will require a much more thorough re-examination of our theories and methodologies for there is an ever present danger that the dualisms will prise apart the connections and associations we might stitch together.' For this reason, we need to be watchful and alert. Whatever we do we should not roll over and go back to sleep. Only the binary dreams of modernism and Western philosophy await us, and not the hybrid world of our waking lives.

Acknowledgement

I would like to thank Ron Johnston for his very helpful comments that improved this chapter.

FURTHER READING

Barnes, T.J. (2003) 'Never mind the economy. Here's culture', in K. Anderson, M. Domosh, S. Pile and N.J. Thrift (eds), *The Handbook to Cultural Geography*. London: Sage. pp. 89–97

Latour, B. (1993) *We Have Never Been Modern*. Cambridge, MA: Harvard University Press. ch. 1

Ray, L. and Sayer, A. (eds) (1999) *Culture and Economy After the Cultural Turn*. London: Sage

Thrift, N.J. and Olds, K. (1996) 'Refiguring the economic in economic geography', *Progress in Human Geography* 20: 31–7

Whatmore, S. (1999) 'Hybrid geographies: rethinking the "human" in human geography', in D. Messey, J. Allen and P. Sarre (eds), *Human Geography Today*. Cambridge: Polity Press. pp. 22–39

References

Amin, A. and Thrift, N.J. (2000) 'What kind of economic theory for what kind of economic geography', *Antipode*, 32: 4–9.

Barnes, T.J. (2001) 'Retheorising economic geography: from the quantitative revolution to the "cultural turn"', *Annals, Association of American Geographers*, 92: 546–65.

Barnes, T.J. (2003) 'The place of locational analysis: a selective and interpretive analysis', *Progress in Human Geography,* 27: 69–95.

Barnes, T.J., Hayter, R. and Hay, E. (2001) 'Stormy weather: cyclones, Harold Innis, and Port Alberni, British Columbia', *Environment and Planning A*, 33: 2127–48.

Bauman, Z. (1989) *Modernity and the Holocaust*. Cambridge: Polity Press.

Beck, U. (1992) *Risk Society: Towards a New Modernity*. London: Sage.

Behrisch, T., Hayter, R. and Barnes, T.J. (2003) 'Resource towns restructuring, youth, and changing labour market expectations: the case of grade 12 students in Powell River, BC', *BC Studies*, 103: 75–103.

Bhabha, H. (1994) *The Location of Culture*. London: Routledge.

Bradley, H. and Fenton, S. (1999) 'Reconciling culture and economy: ways forward in the analyses of ethnicity and gender', in L. Ray and A. Sayer (eds), *Culture and Economy After the Cultural Turn*. London: Sage. pp. 112–34.

Callon, M. (1980) 'Struggles and negotiations to define what is problematic and what is not: the socio-logic of translation', in K.D. Knorr, R. Krohn and R. Whitley (eds), *The Social Process of Scientific Investigation: Sociology of Sciences*, vol. IV. Dordrecht: D. Reidel. pp. 197–220.

Callon, M. (1986) 'Some elements of a sociology of translation: domestication of the scallops and the fishermen of Saint Brieuc Bay', in J. Law (ed.), *Power, Action and Belief: A New Sociology of Knowledge? Sociological Review Monograph* 32. London: Routledge and Kegan Paul. pp. 196–233.

Callon, M. and Law, J. (1995) 'Agency and the hybrid *collectif*', *Southern Atlantic Quarterly*, 94: 483–507.

Chisholm, G.G. (1889) *Handbook of Commercial Geography*. London and New York: Longman, Green, and Co.

Crang, P. (1997) 'Introduction: cultural turns and the (re)constitution of economic geography', in R. Lee and J. Wills (eds), *Geographies of Economies*. London: Arnold. pp. 3–15.

Eagleton, T. (1995) 'Marxism without Marxism', *Radical Philosophy*, 73: 35–7.

Eagleton, T. (2000) *The Idea of Culture*. Oxford: Blackwell.

Edwards, D., Ashmore, M. and Potter, J. (1994) 'Death and furniture: the rhetoric, politics and theology of bottom line arguments against relativism', *History of the Human Sciences*, 8: 25–49.

Egan, B. and Klausen, S. (1998) 'Female in a forest town: The marginalization of women in Port Alberni's economy', *BC Studies*, 118: 5–40.

Gibson-Graham, J.K. (1996) *The End of Capitalism (As We Knew It): A Feminist Critique of Political Economy*. Oxford: Blackwell.

Gibson-Graham, J.K. (2000) 'Poststructural interventions', in E. Sheppard and T.J. Barnes (eds), *A Companion to Economic Geography*. Oxford: Blackwell. pp. 95–109.

Hall, S. (1988) 'Brave new world', *Marxism Today*, October, pp. 24–9.

Haggett, P. (1965) *Locational Analysis in Human Geography*. London: Edward Arnold.

Haraway, D. (1991) *Simians, Cyborgs, Women: The Reinvention of Nature*. London: Routledge.

Haraway, D. (1994) 'A game of cat's cradle: science studies, feminist theory, cultural studies', *Configurations*, 2: 59–71.

Haraway, D.J. (1997) *Modest_Witness@Second_Millenium.Femaleman© Meets_Oncomouse™*. London: Routledge.

Harvey, D. (1982) *The Limits to Capital*. Chicago: Chicago University Press.

Harvey, D. (1989) *The Condition of Postmodernity: An Enquiry into the Origins of Cultural Change*. Oxford: Blackwell.

Harvey, D. (2000) *Spaces of Hope*. Berkeley, CA: University of California.

Hayter, R. and Edgington, D.E. (1997) 'Cutting against the grain: a case study of Macmillan Bloedel's Japan strategy', *Economic Geography*, 73: 187–213.

Johnston, R. (1991) *Geography and Geographers: Anglo-American Geography Since 1945*, 4th edn. London: Arnold.

Krugman, P.R. (1995) *Development, Geography and Economic Theory*. Cambridge, MA: MIT Press.

Lash, S. and Urry, J. (1994) *Economies of Signs and Space*. London: Sage.

Latour, B. (1993) *We Have Never Been Modern*. Cambridge, MA: Harvard University Press.

Latour, B. (1999) *Pandora's Hope: Essays on the Reality of Science Studies*. Cambridge, MA: Harvard University Press.

Law, J. (1994) *Organizing Modernity*. Oxford: Blackwell.

Law, J. (1999) 'After ANT: complexity, naming and topology', in J. Law and J. Hassard (eds), *Actor Network Theory and After*. Oxford: Blackwell. pp. 1–14.

Law, J. (2001) 'Ordering and obduracy', published by the Centre for Science Studies and the Department of Sociology, Lancaster University at http://www.comp.lancs.ac.uk/sociology/soc068jl.html

Leyshon, A. and Thrift, N.J. (1996) *Money/Space*. Oxford: Blackwell.

Lydon, J. (1994) *Rotten: No Irish, No Blacks, No Dogs*. The authorised autobiography of Johnny Rotten of the Sex Pistols, with Keith and Kent Zimmerman. New York: Picador.

Mann, M. (1986) *Sources of Social Power*. Cambridge: Cambridge University Press.

Marcus, G. (1989) *Lipstick Traces: A Secret History of the Twentieth Century*. Cambridge, MA: Harvard University Press.

Martin, R. (1994) 'Economic theory and human geography', in D. Gregory, R. Martin and G. Smith (eds), *Human Geography: Society, Space and Social Science*. London: Macmillan. pp. 21–53.

Martin, R. and Sunley, P. (2001) 'Rethinking the "economic" in economic geography: broadening our vision or losing our focus', *Antipode*, 33: 148–61.

Marx, K. (1904) *A Contribution to the Critique of Political Economy*, translated from the 2nd German edition by N.I. Stone. Chicago: C.H. Kerr.

Massey, D. (1984) *Spatial Divisions of Labour: Social Structures and the Structure of Production*. London: Macmillan.

Mitchell, D. (2000) *Cultural Geography: A Critical Introduction*. Oxford: Blackwell.

Mitchell, K. (1995) 'Flexible circulation in the Pacific Rim: capitalisms in cultural context', *Economic Geography*, 71: 364–82.

Murdoch, J. (1997) 'Inhuman/nonhuman/human: actor-network theory and the prospects for a nondualistic and symmetrical perspective on nature and society', *Environment and Planning D: Society and Space*, 15: 731–56.

Peet, R. (2000) 'Culture, imagery and rationality in regional economic development', *Environment and Planning A*, 32: 1215–34.

Plummer, P. and Sheppard, E. (2001) 'What is the point of economic geography?', *Antipode*, 33: 183–93.

Randall, J.E. and Ironside, R.G. (1996) 'Communities on the edge: an economic geography of resource-dependent communities in Canada', *The Canadian Geographer*, 40: 17–35.

Savage, J. (1993) *England's Dreaming: Anarchy, Sex Pistols, Punk Rock and Beyond*. New York: St Martin's Press.

Sayer, A. (1997) 'The dialectics of culture and economy', in R. Lee and J. Wills (eds), *Geographies of Economies*. London: Arnold. pp. 16–26.

Sheppard, E.S. (1978) Theoretical underpinnings of the gravity hypothesis. *Geographical Analysis*, 10: 386–402.

Smith, J. Russell (1913) *Industrial and Commercial Geography*. New York: Henry Holt and Co.

Smith, N. (1984) *Uneven Development*. Oxford: Blackwell.

Smith, N. (2000) 'What happened to class?', *Environment and Planning A*, 32: 1011–32.

Thrift, N.J. (1999) 'Capitalism's cultural turn', in L. Ray and A. Sayer (eds), *Culture and Economy After the Cultural Turn*. London: Sage. pp. 135–61.

Thrift, N.J. (2000a) 'Pandora's box? Cultural geographies of economies', in G.L. Clark, M.P. Feldman and M.S. Gertler (eds), *The Oxford Handbook of Economic Geography*. Oxford: Oxford University Press. pp. 689–704.

Thrift, N.J. (2000b) 'Performing cultures in the new economy', *Annals, Association of American Geographers*, 91: 674–92.

Thrift, N.J. and Olds, K. (1996) 'Refiguring the economic in economic geography', *Progress in Human Geography*, 20: 31–7.

Whatmore, S. (1999) 'Hybrid geographies: rethinking the "human" in human geography', in D. Massey, J. Allen and P. Sarre (eds), *Human Geography Today*. Cambridge: Polity Press. pp. 22–39.

5 Space: Place

John Agnew

Terrestrial space is often understood as the plane on which events and objects are located at particular places. Consequently, space is general as opposed to the particularity of place (Tuan 1974). Space is also often thought of as commanded or controlled, whereas place is lived or experienced (Taylor 1999). Space is the abstraction of places into a grid or coordinate system as if the observer or controller is outside of or looking down on the places that constitute it. In much academic usage, however, in human geography and in other fields, space and place are either not distinguished at all, but viewed as synonymous, or one is seen as trumping the other. In this latter view, different intellectual traditions tend to express themselves in the preference for one word or the other, in part simply to distinguish themselves through association with one or the other but also because the terms are seen as representing totally different views of the nature of the world and geographical representations of it.

This polarization of positions on space and place is important in recent geographic thought, space being associated with objectivist theories, as in spatial analysis and Marxist political economy, and place with subjectivist theories, as in phenomenology and postmodernism. As a result, the two terms are increasingly seen as representing alternative conceptions of spatiality (how space/place enter into human lives and social arrangements) rather than as inherently or internally related to each other. This increasingly dominant conception of space and place follows a long period in which the two terms, often blended, were also inadequately theorized. Yet, as the initial definition given above suggests, each needs to be seen as invariably depending on the other for either to have much by way of a satisfactory theoretical basis. Indeed, I would claim that contemporary usage of both terms suffers from a tendency to either overextend one at the expense of the other often by failing to mention the other or by associating it with a philosophical or political position that is totally opposed to that of the writer in question. The argument of this chapter is that adequate understanding of either term requires that each be related theoretically to the other.

Space and place have experienced something of an efflorescence as theoretical terms in recent years, partly because they have been discovered by writers in anthropological, sociological, literary and post-colonial studies in search of an exit from the theoretical cul-de-sac of presuming national territories or other self-evident spatial units such as the 'urban' or the 'global' exhaust the application of spatiality (for example, Dirlik 2001; Jiménez 2003). But it is also undoubtedly because human geography has become increasingly focused on the concepts that define the field's distinctiveness and contribution to wider intellectual debates. Different schools of thought expropriate different concepts and then imbue them with meanings that facilitate the larger theoretical goals of the school in question (for example, Harvey 1993). There is a need to reconnect the two concepts, space and place. As conceptual twins, they offer more together than use of either does separately. Only when related to the other does either reach its full potential.

In this chapter I begin by providing basic or 'common sense' definitions of the two terms drawn from a range of writers and why they have come to be seen increasingly as opposites. I then turn to their recent histories, paying special attention to the fusion with or exclusion of the other by proponents of each. In a fourth section I argue for how the two terms are both distinctive yet mutually dependent and how this helps in understanding various phenomena from uneven economic development and cultural differences to 'how we know'. I review some of the ways in which the two are being related to each other. A short conclusion summarizes the main points and suggests that each necessitates the other despite repeated attempts to champion one or overcome one or the other.

Defining space and place for geographical studies

In the simplest sense, space refers to location somewhere and place to the occupation of that location. Space is about having an address and place is about living at that address. Sometimes this distinction is pushed further to separate the physical place from the phenomenal space in which the place is located. Thus, place becomes a particular or lived space. Space then refers to the fact that places cannot but be located somewhere. Place is specific and space is general.

So far, so good. By and large, what I have written above is pretty much uncontroversial. This basic formulation, however, has been subject to two very important but controversial redefinitions down the years. The first is that the language of space and place is often elided with the language of geographical scale. Though there is no necessary connection of this type, the usage is very common, with place standing in for the local (and traditional) and space representing the global (and the modern) (for example, Harvey 1989). This conflation draws attention to a further feature of how the two terms are frequently used. Place is often associated with the world of the past and space with the world of the present and future. From one perspective, place is therefore nostalgic, regressive or even reactionary, and space is progressive and radical. From another side, place is being lost to an increasingly homogeneous and alienating sameness. Placelessness is conquering place as modernity displaces traditional 'folkways'. From both perspectives, usage reflects a subtle incorporation of time into how the terms are defined. Place is the setting for social rootedness and landscape continuity. Space represents the transcending of the past by overcoming the rootedness of social relations and landscape in place. Not surprisingly, from this understanding those on the political Left will tend to hope that space does trump place, in the hope of a less unequal and more secular world. The case for place, on the other hand, has often been championed by nationalists, conservatives and believers in the importance of sacred places. Presumably, therefore, the political Right should champion place over space. Yet, each of these attributions is mistaken. Neither term has an inherent political association.

Secondly, the priority given to either space or place (and the fusion with or exclusion of the other) depends on what is made of the nature of space. In particular, what is made of the distinction rests in large part on whether Newton's or Leibniz's understanding of space prevails (Casati and Varzi 1999: 21). Discussion of space, as opposed to place, is a modern concern. Indeed, it could be seen as one of the markers of modernity, dating at most from seventeenth-century Europe, and associated with such intellectual giants as Descartes, Newton, Leibniz and Kant. From one point of view, argued persuasively by Casey (1993: 8), place became subordinated to space (and both to time) in the seventeenth century and has only been tentatively rehabilitated in the twentieth. The project of a 'spatial history' that can be associated with Heidegger and Foucault depends fundamentally on relating place to space as if they are internally related to each other (Elden 2001: 90–1). Much contemporary understanding of space and place, however, seems to depend on relatively unreformed seventeenth-century conceptions of what they mean, specifically Newtonian and Leibnizian ones.

In the Newtonian view, space is absolute, in the sense that it is an entity in itself, independent of whatever objects and events occupy it, containing these objects and events, and having separate powers from them. Newton's view is best used to describe the motion of rigid bodies through 'empty' space. Newton was a transitional figure in the genesis of modernity, so it is not surprising that he would endorse a medieval or thing-like conception of space. In other words he conceived of absolute space as previous generations of thinkers had always thought of particular places without recourse to modern ideas of space at all: as concrete and real. In the Leibnizian view, space is relational, in the sense that it has no powers independent of objects and events but can be construed only from the relations between them. In other words, Leibniz is the founder of the modernist view of space as that which gives it no powers in itself. Only within a Newtonian view can space be 'active' in itself. But this is because space is made up of places. In a Leibnizian view it is the powers of events and objects taking place that make space appear 'active'. But space is entirely parasitic on the relations between objects and events occupying places. Space thus exists because of relations between sites at which events and objects are located. In both cases, of course, neither space nor place can exist without the other. But the priorities obviously differ. As a result, subjectivists, emphasizing the subjective orientation of human actors in places that condition them, are usually Newtonian in their understanding of space and emphasize place, whereas objectivists, emphasizing the role of causal 'forces' in human life, tend to a Leibnizian position and refer to relative space and spatial relations. This association often seems to lead to the 'other' concept receiving short shrift when a social scientist becomes committed primarily to one or the other, irrespective of whether the thinker is either a self-conscious Newtonian or Leibnizian. In the history of thinking about space, it is only with Einstein's space–time that space is finally seen as *interdependent* with the distribution of things or matter (Stein 2002: 292). But much thinking about space in the social sciences is still resolutely pre-Einsteinian.

Of course, you could say that, as long as thinkers are clear about their philosophical and political orientations, how they use the terms space and place is unproblematic. For example, if de Certeau (1984) reverses the usual meanings of the two words so that space is thought of as activated places and place is the field of relations in which spaces are embedded, then there is not a problem so long as readers know what has happened. But the conflation of space and place with temporal/scale political interpretations and the adoption of either purely absolutist or relational views of space (usually by analogy to physical mechanics) tend to have both negative intellectual and political consequences. They also load the deck against seeing space and place as internally related to one another. According to Newtonians, space and place are fixed as eternal opposites, while for the followers of Leibniz, space is either the direct effect of the placement of objects and events on one another or simply the sites of events and objects without any separate 'effect'. If space and place are taken as internally related to one another in human practice, however, these oppositions dissolve. Space then signifies a field of practice or area in which a group or organization, such as a state, operates, held together in popular consciousness by a map-image and a narrative that represents it as a meaningful whole. Place represents the encounter of people with other people and things in space. It refers to how everyday life is inscribed in space and takes on meaning for people and organizations. Space is thus 'top-down', defined by powerful actors imposing their control and narratives on others. Place is the 'bottom-up', representing the outlooks and actions of more typical folk. Places can be localized when associated with the familiar. But they can also be larger areas, depending upon patterns of activities, network connections and the projection of feelings of attachment, comfort and belonging. From this perspective, neither trumps the other. Indeed, each requires the other (Agnew and Smith 2002).

Space over place

Among some geographers in the 1960s, place (in the form of the analogous concept of region) was associated with the *ancien régime* of geography, and space in itself was to be the new object of study. Giving an explicit privilege to space over place, then, goes back at least to the 'spatial revolution' in geography in the 1960s but it is also apparent among those who for political reasons associate place negatively with the past and space positively with the future. Spatial analysis may have failed to capture the field in its entirety but it lives on in the practices of many geographers committed to mapping spatial patterns and reasoning back to the social, economic and/or political processes that may have produced them. The focus is on relationships between events and objects in space by correlating their spatial co-occurrence. In a simple but classic example, Arthur Robinson (1962) mapped and graphed the correspondence between rural farm population density and annual precipitation for the state of Nebraska showing a strong relationship between the two. Though not a perfect 'match' by any means, the strong relationship suggests a possible causal relationship between precipitation and population density in an agricultural area. This sort of example was used by pioneers in spatial analysis, such as Abler, Adams and Gould (1971: 56) to argue that the 'distinctively geographical question is "Why are spatial distributions structured the way they are?"' The term place is used, but only as a synonym for location or point. A place exists only as a node on a spatial surface that can be classified or related to other places in terms of the objects and events that are the primary object of analysis.

This approach was challenged in the 1970s by those who claimed that the nature of space and its effects depends on the economic conditions under which space is 'produced'. Initially, the focus was on the ways in which economic processes, particularly the establishment of ground rents and the search for profit, produce spatial patterns of investment and divestment. Later, however, Marxist geography and other approaches drawing from critical social theory came to argue that the arrival of capitalism has combined in dialectical tension a materialist relativizing of space (making the whole world a production surface) with an ideological absolutizing of space (dividing global space into states and empires, clear demarcations of property rights in land, etc.) (for example, Smith and Katz 1993: 75). There is nothing 'bottom-up' about any of this. It is a strictly 'top-down' world. One particular strand of thought, that associated with the world-systems analysis of Immanuel Wallerstein, divides the world geographically into three zones (core, semiperiphery and periphery) in which different combinations of economic and political processes prevail because of the overall working of the modern world economy. The world economy operates to produce flows of resources, goods and profits from the periphery to the core. In this relative space elites attempt to construct absolute (state) spaces within which they can have control. They eventually always fail because of the overwhelming force exerted by the world economy in controlling and channelling economic development. From this point of view, illustrated in papers by Taylor and others (for example, Osei-Kwame and Taylor 1984 on politics in Ghana), what happens at any place is a direct result of the zone in which it is located. Thus, to take the example of politics in Ghana, politics there is held to revolve around competition between two political elites: that from the cocoa-growing region, which favours relatively free trade, and that representing parts of the country that do not profit from the cocoa business and which would like to see greater economic protectionism. These different places within the country have no constitutive role themselves in Ghanaian politics. They just happen to be sites in global space where certain determinative core–periphery relationships are played out.

But place is not always simply overwhelmed by the presumptive powers of spatial relations (either natural or socially constructed). In other words, it is not always just a question of fusion but sometimes a question of explicit exclusion. Quite openly, place is often decried or condemned as retrograde, reactionary and/or misleading. Particularly given to the exclusion of place are perspectives with roots in teleological and evolutionary conceptions of human history. Prominent among them are modernization and neo-Marxist theories that see human history as an automatic movement from traditional to modern or from capitalism to socialism. Thus Michael Hardt, a literary critic, and Antonio Negri, a neo-Marxist philosopher (Hardt and Negri 2000: 40–5, 208–10), co-authors of a widely circulating book, define modern global space as an 'Empire' in which 'Today nearly all of humanity is to some degree absorbed within or subordinated to the networks of capitalist exploitation' (p. 43). These global networks should be the basis for organizing a politics of 'resistance', not 'place-based' movements in which 'the boundaries of place (conceived either as an identity or a territory) are posed against the undifferentiated and homogeneous space of global networks' (p. 44). Place is thus associated with 'localism' and 'locality'. Later on the authors suggest why they think this way. Today, they aver, 'The dialectic between productive forces and the system of domination no longer has a determinate *place*' (p. 209; original emphasis). Place is dead. All we now have is Empire. This is a conception of space as a giant global pinhead. Notwithstanding the ontological difficulty of imagining political activity taking place globally without any nodes (or places) of direction, it seems both hyperbolic and problematic to make this sort of claim at a time when, to choose just two empirical examples from a potentially much larger field of examples, large parts of Africa have fallen out of the world economy and people are fighting to the death around Jerusalem for control over a holy site. But it reflects a common outlook, and joins the putatively radical Hardt and Negri with neo-liberal proponents of a 'world that knows no boundaries', in which place may once have had importance (at least as the location of something, if not as a unique setting) but, as a result of the rise of modernity and/or capitalism, has been eclipsed. Now there is no need for place whatsoever. Indeed, it must be excluded for fear that opponents of Empire will focus on defending (fictive) places against the inroads of (real) global space. Consequently, in this rendering of the absolute victory of space over place you must forget about place or be rendered politically impotent.

Place over space

On the other side of the coin, so to speak, are those who focus on the first-hand experience of human subjects in place. Space is bracketed, or put to one side, because its 'abstractness discourages experiential explorations' (Casey 2001: 683). In his philosophical rehabilitation of place, Casey (1997, x) notes how 'place has been assimilated to space. ... As a result, place came to be considered a mere "modification" of space (in Locke's revealing term) – a modification that aptly can be called "site," that is, leveled-down, monotonous space for building and other human enterprises'. Casey's goal is to argue for the crucial importance of place in much thinking about community and the public sphere, even though the connections are often not made explicit. He wants to make place different from site and space, even though he acknowledges Michel Foucault's point that the modern world is largely one of Leibnizian sites and relations (Casey 1997: 298–300). In rethinking space as place, his primary interest lies in phenomenologically linking places to selves (also see Entrikin 1991, 2001). The central issue is that of 'being in place differently' (Casey 1997: 337) conditioning the various dimensions of selfhood, from the bodily to the psychological, institutional and architectural. So, though the 'shape' of place has changed historically, it is now no mere container but, rather, a taking place, its rediscovery and naming as such is long overdue. Thus: 'Despite the seduction of endless space (and the allure of serial time), place is beginning to escape from its entombment in the cultural and philosophical underworld of the modern West' (Casey 1997: 339).

According to Casey, and others, evoking the 'divine' or 'sacred' in everyday places is perhaps one of the most instructive examples of the recent revival or revaluation of place. Associated at one time primarily with holy sites (ancient temples, battlefields, holy cities, etc.), the sacred place is seen as representing the emotional experience of unity and order in a world made up predominantly of profane and chaotic space (Eliade 1959; Kuper 1972). If in the great monotheistic religions the divine is usually characterized as co-extensive with infinite space, however, it is in places that it is revealed. Consequently, the divine can be manifested in the places people inhabit in everyday life. So, sacred places are not just sites of pilgrimage or of religious veneration. The vocabulary now is therefore of the particular, the vernacular and the personal in ordinary places. Theologians and sociologists of the everyday have discovered place. One such commentator expresses his interest in place as a concept for understanding the sacred in the everyday in the following way, noting the ascriptive power of place more than the role of personal agency: 'To be a person is not merely to be embodied but also to inhabit a public place. Our social selves are created for us, not just symbolically but also physically within roles determined by social, cultural and religious hierarchies as well as gender stereotypes' (Sheldrake 2001: 21). All of us inhabit sites in such places, even when we are exiles, displaced people, or members of diaspora communities. A major tension exists between trying to claim a presence in a place or leaving such a place behind. This is a long-standing dilemma, as anyone who has ever moved can attest to. One of the great political challenges in the contemporary world concerns how to open up belonging to places for those currently marginalized and oppressed within them. The immigration 'problem' and the 'problem' of extending citizenship and membership rights to excluded groups, therefore, become problems of gaining access to place. More generally, under conditions of globalization it is space that loses its grip, particularly in the form of the state territory, and opens up the possibility for a renewed 'politics of presence' in what Adriana Cavarero (2002: 525), drawing from Hannah Arendt, calls 'locality without territory'.

To some writers, however, place is under threat from a generalized condition of placelessness. This is not seen as the victory of space over place. Rather, it is the diminution in the capacity of place to provide the 'goods' that it once did. The local specificity and variety of places have been replaced by relatively standardized and thus increasingly homogeneous landscapes. Here again, the issue is not that of a fusion of space with place (as with Casey and Sheldrake) or the victory of place over space (as with Cavarero) but of exclusion; in this case, of any attention to space whatsoever. To Edward Relph (1976), for example, the loss of diversity among places is symptomatic of a wider loss of meaning. The 'authenticity' of pre-industrial places has been lost because of a creeping 'inauthentic' attitude that values sameness more than difference. Relph points to such examples as strip malls, new towns, the international style in architecture and tourist landscapes. Of course, these changes could be seen in a very different light. For one thing, though some meanings are undoubtedly lost, others are also gained. Placelessness is not synonymous with meaninglessness. For another, the idea of a single sense of place may be both romantic and reactionary. Any particular place can be many different places to different people. Place is not the same as an organic community existing in isolation, cut down by a sweeping scythe of placelessness (Oakes 1997). 'Placelessness' lies in the eye of the beholder.

Putting space and place together

Symptomatic of the conceptual separation of space and place are the three dominant meanings that geographical place has acquired in writing that invokes either space or place (Agnew 1987, 1989, 1993). Each meaning tends to assimilate place to one or the other end of a continuum running from nomothetic (generalized) space at one end to idiographic (particularistic) place at the other. The first is place as location or a site in space where an activity or object is located and which relates to other sites or locations because of interaction and movement between them. A city or other settlement is often thought of this way. Somewhere in between, and second, is the view of place as locale or setting where everyday-life activities take place. Here the location is no mere address but the where of social life and environmental transformation. Examples would be such settings from everyday life as workplaces, homes, shopping malls, churches, etc. The third is place as sense of place or identification with a place as a unique community, landscape and moral order. In this construction, every place is particular and, thus, singular. A strong sense of 'belonging' to a place, either consciously or as shown through everyday behaviour such as participating in place-related affairs, would be indicative of 'sense of place'.

Attempts at putting space and place together must necessarily try to bring at least two of these various meanings of geographical place together. Currently, there are four main ways in which this task has been approached: the humanist or agency-based, the neo-Marxist, the feminist and the contextualist-performative. Each of these rejects the either/or logic in relation to space and place that has characterized most geographic and social thought from the sixteenth century to the present. For the first, the focus lies in relating location and locale to sense of place through the experiences of human beings as agents. In one of the most sophisticated statements of this perspective, Robert Sack (1997: 58) provides the essential thrust when he writes that his 'framework draws on the geographical experiences of place, space, home and world which people use in their lives to integrate forces, perspectives, and selves'. From this point of view:

> Place implies space, and each home is a place in space. Space is a property of the natural world, but it can be experienced. From the perspective of experience, place differs from space in terms of familiarity and time. A place requires human agency, is something that may take time to know, and a home especially so. As we move along the earth we pass from one place to another. But if we move quickly the places blur; we lose track of their qualities, and they may coalesce into the sense that we are moving through space. This can happen even in my own home. If I am hardly there and do not attend to its contents, it may seem unfamiliar to me, more like a part of space than a place. (Sack 1997: 16)

In this frame of reference, cultural differences, for example, emerge because of place-based experiences and human agency but also because places are never separate but always part of larger sets of places across which differences are more or less pronounced depending on the permeability of boundaries between places as people experience them. Places are woven together through space by movement and by the network ties that produce places as changing constellations of human commitments, capacities and strategies. Places are invariably parts of spaces and spaces provide the resources and the frames of reference in which places are made.

The neo-Marxist perspective on relating space and place is best represented in the writings of Henri Lefebvre (1991) and his geographic interpreters (for example, Harvey 1989; Soja 1989; Merrifield 1993). This approach focuses on location and locale as the two central features of place in relation to space. In his writing, Lefebvre has focused on the social production of the spaces within which social life takes place. Under capitalism, 'abstract space' (the space produced by economic transactions and state policies) has 'colonized' everyday life by means of both spatial practices (commodification and bureaucratization) and representations of space (discourses of planning and surveillance). Lefebvre looked for a movement against this colonization of concrete space (or place) to reclaim the spaces of everyday life. This could be accomplished by insurgent 'counter-discourses' based on spaces of representation that build on memories and residues of an older 'authentic' existence and new practices in concrete space. Merrifield (1993: 521, 525), for example, interprets Lefebvre as seeing space and place as dialectically related such that space is a 'rootless, fluid reality of material flows' or 'the realm of dispassionate "objects" rationally "ordered in space"', that Lefebvre called the 'realm of the conceived', whereas place 'comprises the locus and a sort of stopping of these flows', what Lefebvre called the 'realm of the lived'. But this account seems to understate the degree to which 'concrete space' (or place) for Lefebvre signifies a 'bottom-up' and autonomous reaction to the depredations of those agents of capital and the state whose dominance has produced abstract space. The use of the phrase 'dialectically related' obfuscates rather than clarifies quite what the relationship is between abstract space and concrete place. Nevertheless, this framework does suggest how uneven economic development is jointly produced by dominant practices and discourses but can only be challenged by and on behalf of people in places attempting to recapture concrete (place) space from the abstract space of modern capitalism (also see Taylor 1999).

The postmodern-feminist perspective, broadly construed, is one that is both deeply suspicious of grand narratives that suppress multiplicity, difference and alterity in their obsession with temporal transitions (from tradition to modern, etc.) and sympathetic to rethinking space in terms of multiplicity and dislocation. In this framework, 'places may be thought of as open articulations of connections' and 'identities of subjects and identities of places constructed through interrelations not only challenge notions of past authenticities but also hold open the possibility of change in the future' (Massey 1999: 288). Place, then, includes location, locale and sense of place but without the central focus on human agency that brings these together in the humanist perspective. Here the emphasis is on places as sites in the flow of social relations. Place is seen as constituted out of space-spanning relationships, place-specific social forms and a sense of place associated with the relative well-being, disruption and experience of living somewhere. The experience of place is quite different for different groups, such as children, women, subordinated social classes, minorities, etc. Feminism claims that many accounts insist on denying this pluralism. In particular, the argument goes, they also privilege space by associating it with 'high politics' and 'Reason' and denigrate place as personified by home and the mundane lives of women and children (Massey 1994). Reasserting place, therefore, is a political as much as an intellectual move. It is about re-valorizing place not as the site of domesticity but as a meeting place. As Doreen Massey (1994: 154), the most articulate advocate of this viewpoint, suggests:

> Instead then, of thinking of places as areas with boundaries around, they can be imagined as articulated moments in networks of social relations and understandings, but where a large proportion of those relations, experiences and understandings, are constructed on a far larger scale than we happen to define for that moment as the place itself, whether that be a street, or a region or even a continent. And this in turn allows a sense of place which is extroverted, which includes a consciousness of its links with the wider world, which integrates in a positive way the global and the local.

The very possibility of political agency, understood as always collective or group-based, depends upon this 'open' and unbounded conception of place. From this perspective, other conceptualizations, particularly those that privilege space over place, tend to leave little likelihood of political change coming from bottom-up political organization.

The contextualist-performative perspective is also suspicious of grand theories but also finds problematic the division between representation and practice that most attempts at relating space and place, particularly the neo-Marxist, tend to take for granted (for example, Barad 2003). In this view, as Nigel Thrift (1999: 317) avers, place is associational, weaving together 'all manner of spaces and times', but it is never completed because it always depends on 'further works of association'. Drawing on post-humanist actor-network theory, but critical of its reluctance to engage with the 'role of common ground' or place in 'how networks echo back and forth', Thrift (1999: 313) emphasizes the materiality of places as 'open spaces' that practices make and that take 'shape only in their passing' (Thrift 1999: 310). In this view, the particularities of any situation cannot be read off from the predictions of a grand theory. Rather, places are specific time–space configurations made up of the intersection of many encounters between 'actants' (people and things) that reflect '*practical means of going on rather than something concerned with enabling us to see, contemplatively, the supposedly true nature of what something is*' (Thrift 1999: 304; original emphasis). This viewpoint, therefore, is as much epistemological (concerned with how we know) as ontological (concerned with what exists). We always look at 'the world' from somewhere, from a place. So, knowledge is always geographically contextual and reflexive. Space, then, is always and everywhere implicated in what people (and other entities) do and how they think, not just where they are. 'Space is no longer a category of fixed and given ontological attributes, but a becoming, an emerging property of social relationships' (Jiménez 2003: 140). This position is thus much the most radical of the four in attempting to relate space and place together by raising the difficulty of ever thinking beyond the confines of the space–time context in which we are embedded and the fact that there is spatiality to social relationships even when 'it is no longer necessary to download meaning onto a territory [place] or weave it into land' (Jiménez 2003: 150). In highlighting these themes, however, the position comes close to endorsing a geographical relativism that would make its own claims to say something general about place suspect!

Each of these positions offers a potential route towards recognizing and articulating the inherent relationship of space to place. Of course, each emerges out of distinctive intellectual traditions and thus cannot be seen as a substitute for any one of the others on the road to overturning the dominance of either/or thinking about space and place. In other words, each has its own problems and potential. If I personally prefer the first as apparently the most convincing, it is only because I accept that position's emphasis on human agency as much the best way of bringing the three dimensions of place together within one coherent theoretical framework. Its tendency to confuse human agency with individual personal autonomy or the asocial self, however, needs close critical scrutiny (Agnew 2002).

Concluding remarks

In many contemporary accounts, space and place are radically separated from one another either through a strategy of fusion (folding one into the other) or of exclusion (denigrating or devaluing the other). I suggested that the elision of the language of space and place with that of geographical scale (space = global, place = local) and the negative impact of Newtonian and Leibnizian conceptions of space bear much of the responsibility for the polarized use of the two terms. The burden of my argument has been that this separation is intellectually and politically untenable. Each requires the other to fulfil its potential. To do so, however, means adopting a non-modernist understanding of space and place, to use Steven Moore's (2001) word: one that rejects the either/or logic that pervades most modernist (and postmodernist) thinking about space and place. I have briefly outlined four ways in which space and place can be distinguished from one another yet maintain a role in relation to one another. Each of these requires abandoning currently dominant conceptions of space and place but hopefully for perspectives that are both intellectually and politically more relevant to the times.

FURTHER READING

Agnew, J.A. (2002) *Place and Politics in Modern Italy.* Chicago: University of Chicago Press

Understands place as the result of a web of spatial-temporal connections in which political agency is embedded. Offers empirical case studies of Italian unification and recent politics to argue for the superiority of this perspective over others that treat Italy as a single, homogeneous space or a set of fixed regions with long-standing cultural traits that determine political outcomes.

Massey, D. (1994) *Space, Place, and Gender.* Minneapolis: University of Minnesota Press

A series of essays clearly reviewing prominent positions on space and place and offering a forceful statement of Massey's feminist-socialist perspective on political agency and a 'global sense of place'. Particularly important are Chapters 6 ('A global sense of place'), 7 ('A place called home'), 10 ('Flexible sexism') and 11 ('Politics and space/time').

Merrifield, A. (1993) Place and space: a Lefebvrian reconciliation. *Transactions of the Institute of British Geographers NS* 18: 516–31

An accessible paper showing how the ideas of the French Marxist philosopher Henri Lefebvre can be used to reconcile between 'the way in which experience is lived and acted out in place' and 'the political and economic practices that are operative over broader spatial scales' (p. 517). Attempting to abandon the neo-Kantian nomothetic-idiographic opposition around which most discussion of space and place has tended to revolve, Merrifield argues for a 'dialectical' perspective on relations between what the two terms signify. In this view, space and place are 'two facets of a dialectical process just as the wave and particle aspect of matter is assumed in particle physics' (p. 527). In other words, space and place are two 'moments' in a common process with space equivalent to the 'whole' and place to the 'parts'.

Sack, R.D. (1997) *Homo Geographicus.* Baltimore: Johns Hopkins University Press

A sophisticated theory of how space and place are related from a broadly humanist perspective, emphasizing the roles of human awareness and moral concern as well as more conventional social and environmental dimensions of experience. Places are both what tie together movements through space and dense nodes of relations between people and with nature. As geographical agents, people make places but not entirely as they please.

Taylor, P.J. (1999) Places, spaces and Macy's: place–space tensions in the political geography of modernities. *Progress in Human Geography*, 23: 7–26

An interesting and accessible argument for why space and place are inherently related to one another through the 'tension' implicit in their dominant meanings. If space is the outcome of 'top-down' processes, place is the result of 'bottom-up' ones. After claiming that space and place are the most basic of geographic concepts, Taylor argues for absolutely the reverse of Hardt and Negri's (2000) claim that progressive politics requires abandoning place-based organization for the nebulous spaces of globalized networks. To Taylor, space is not only more 'abstract' than place it is also necessarily imposed rather than experiential and popular in origin. Thus, through understanding 'space–place tensions' a progressive politics can provide both a perspective on state- and economic-based forces manipulating space and a place-based response to these depredations. From this point of view, a space-based politics is necessarily authoritarian.

Thrift, N. (1999) Steps to an ecology of place. In D. Massey, J. Allen, and P. Sarre (eds) *Human Geography Today*. Cambridge: Polity Press

An accessible account of 'non-representational' approaches to space and place emphasizing Wittgenstein's and Heidegger's contributions to ideas that conceive of place-making as 'dwelling' (the practices of everyday) rather than 'building' (planning and then constructing). The performative and practical are given central billing over the representational and determinative.

Tuan, Y-F. (1974) Space and place: humanistic perspective. *Progress in Geography* 6: 211–52

The 'classic' statement of the inevitable duality of space and place, emphasizing what each has to offer separately but committed to relating them one to another. At the time of its writing, this article expressed a sentiment that was definitely in short supply and was an important reminder of the danger of traditions that would either define the two words in entirely polarized and oppositional terms or fuse them unintelligibly.

References

Abler, R., Adams, J.S. and Gould, P. (1971) *Spatial Organization: The Geographer's View of the World*. Englewood Cliffs, NJ: Prentice-Hall.

Agnew, J.A. (1987) *Place and Politics: The Geographical Mediation of State and Society*. London: Allen and Unwin.

Agnew, J.A. (1989) The devaluation of place in social science. In J. Agnew and J. Duncan (eds), *The Power of Place: Bringing Together Geographical and Sociological Imaginations*. London: Unwin Hyman.

Agnew, J.A. (1993) Representing space: space, scale and culture in social science. In J. Duncan and D. Ley (eds), *Place/Culture/Representation*. London: Routledge.

Agnew, J.A. (2002) *Place and Politics in Modern Italy*. Chicago: University of Chicago Press.

Agnew, J.A. and Smith, J.M. (eds) (2002) *American Space/American Place: Geographies of the Contemporary United States*. Edinburgh: Edinburgh University Press.

Barad, K. (2003) Posthumanist performativity: towards an understanding of how matter comes to matter. *Signs* 28 (3): 801–31.

Casati, R. and Varzi, A.C. (1999) *Parts and Places: The Structures of Spatial Representation*. Cambridge, MA: MIT Press.

Casey, E.S. (1993) *Getting Back into Place: Toward a Renewed Understanding of the Place-World*. Bloomington, IN: Indiana University Press.

Casey, E.S. (1997) *The Fate of Place: A Philosophical History*. Berkeley and Los Angeles, CA: University of California Press.

Casey, E.S. (2001) Between geography and philosophy: what does it mean to be in the place-world? *Annals of the Association of American Geographers* 91 (4): 683–93.

Cavarero, A. (2002) Politicizing theory. *Political Theory* 30 (4): 506–32.

de Certeau, M. (1984) *The Practice of Everyday Life* (translated by S. Rendell). Berkeley and Los Angeles, CA: University of California Press.

Dirlik, A. (2001) Place-based imagination: globalism and the politics of place. In R. Prazniak and A. Dirlik (eds), *Places and Politics in an Age of Globalization*. Lanham, MD: Rowman and Littlefield.

Elden, S. (2001) *Mapping the Present: Heidegger, Foucault and the Project of a Spatial History*. London: Continuum.

Eliade, M. (1959) *The Sacred and the Profane*. New York: Harcourt Brace.

Entrikin, J.N. (1991) *The Betweenness of Place: Towards a Geography of Modernity*. Baltimore, MD: Johns Hopkins University Press.

Entrikin, J.N. (2001) Hiding places. *Annals of the Association of American Geographers* 91 (4): 694–7.

Hardt, M. and Negri, A. (2000) *Empire*. Cambridge, MA: Harvard University Press.

Harvey, D. (1989) *The Condition of Postmodernity*. Oxford: Blackwell.

Harvey, D. (1993) From space to place and back again: reflections on the condition of postmodernity. In J. Bird et al. (eds), *Mapping the Futures: Local Cultures, Global Change*. London: Routledge.

Jiménez, A.C. (2003) On space as capacity. *Journal of the Royal Anthropological Institute* 9 (1): 137–53.

Kuper, H. (1972) The language of sites in the politics of space. *American Anthropologist* 74: 411–25.

Lefebvre, H. (1991) *The Production of Space*. Oxford: Blackwell.

Massey, D. (1994) *Space, Place, and Gender*. Minneapolis, MN: University of Minnesota Press.

Massey, D. (1999) Spaces of politics. In D. Massey, J. Allen and P. Sarre (eds), *Human Geography Today*. Cambridge: Polity.

Merrifield, A. (1993) Place and space: a Lefebvrian reconciliation. *Transactions of the Institute of British Geographers NS* 18: 516–31.

Moore, S. (2001) *Technology and Place*. Austin, TX: University of Texas Press.

Oakes, T. (1997) Place and the paradox of modernity. *Annals of the Association of American Geographers* 87 (3): 509–31.

Osei-Kwame, P. and Taylor, P.J. (1984) A politics of failure: the political geography of Ghanaian elections, 1954–1979. *Annals of the Association of American Geographers* 74 (4): 574–89.

Relph, E. (1976) *Place and Placelessness*. London: Pion.

Robinson, A.H. (1962) Mapping the correspondence of isarithmic maps. *Annals of the Association of American Geographers* 52: 414–29.

Sack, R.D. (1997) *Homo Geographicus*. Baltimore, MD: Johns Hopkins University Press.

Sheldrake, P. (2001) *Spaces for the Sacred: Place, Memory, and Identity*. Baltimore, MD: Johns Hopkins University Press.

Smith, N. and Katz, C. (1993) Grounding metaphor: towards a spatialized politics. In M. Keith and S. Pile (eds), *Place and the Politics of Culture*. London: Routledge.

Soja, E. (1989) *Postmodern Geographies: The Reassertion of Space in Critical Social Theory*. London: Verso.

Stein, H. (2002) Newton's metaphysics. In I.B. Cohen and G.E. Smith (eds), *The Cambridge Companion to Newton*. Cambridge: Cambridge University Press.

Taylor, P.J. (1999) Places, spaces and Macy's: place–space tensions in the political geography of modernities. *Progress in Human Geography* 23: 7–26.

Thrift, N. (1999) Steps to an ecology of place. In D. Massey, J. Allen, and P. Sarre (eds), *Human Geography Today*. Cambridge: Polity.

Tuan, Y-F. (1974) Space and place: humanistic perspective. *Progress in Geography* 6: 211–52.

6 Black:White

Susan J. Smith

Thinking in dualisms has become highly unpopular. The late twentieth century was a period when hybridity and in-betweenness, flexibility and mutability gained conceptual currency over fixity, stability and boundedness in accounting for patterns of social life. Almost everything written on the topic regards this newfound fluidity in thought, and in fact, to be a sign of intellectual progress; there is often a sense that what went before is either dangerously or laughably flawed.

What I want to offer in this chapter is a slightly different version of history. It is a history whose integrity is irretrievably damaged by the tenacity of powerful racial ideologies. These are ideologies whose efforts to distinguish, differentiate and distance have, from generation to generation, legitimized a catalogue of atrocities perpetrated by humans against humans, on the grounds that some are different from, and inferior to, others. In revisiting this history, it is easy to see the attraction of an increasingly 'hybrid' world that can put an end to the idea of race, and to the racisms that go with it. However, what this history also shows is that working with a black/white binary (and, increasingly, with other race categories indexed by phenotype) has been, and to an extent remains, an important anti-racist strategy. Working without such categories, moreover, raises issues that the research community has scarcely begun to imagine.

The starting point for this chapter is the old and erroneous, yet enduring, idea that human populations can be subdivided into racial types on the basis of traits which are written on the body – skin colour, physique, facial features – which are taken to signal a common point of origin (a homeland, and/or a particular family history or ancestry), used to index a range of social practices (language, religion, traditions), and taken-for-granted as an axis of inequality. Of course, articulated in this way the idea seems crude and outdated. How people identify themselves, how they express their shared meanings and cultures is one thing – but 'race' is a matter of how people are categorized; and grouping people, and dispensing resources and life chances among them, according to physical, behavioural, cultural and emotional traits that are assumed to be inherent or inherited is, we know, a (racist) social and political practice.

However, it is precisely to establish this point that the binary black/white became such a powerful signifier in debates on race and racism. Indeed, the core of my argument is that categorical thinking about human races – indexed above all by the binary formulations black/white and white/other – was a strategic response on the part of liberal and radical thinkers to the social harm posed, in the late nineteenth and early twentieth centuries, by a more pervasive and politically conservative dualism, namely that which divides culture and society on the one hand from nature on the other. The nexus of culture–nature–race is considered in the first section of this chapter.

In the second part of the chapter, I go on to show that arguments in favour of the social, political and economic construction of race categories were set up as an explicitly anti-racist strategy. This way of thinking mounted a direct challenge to prevailing wisdoms, which favoured the naturalization of racialized inequality. Thinking in binary – usually 'black', but sometimes 'non-white', 'brown', or 'of colour', and 'white' – was a way of recognizing that what was really at issue in the race debate was the appropriation and exploitation of the material wealth, cultural life and psychic well-being of (non-white) colonized peoples by (white) colonizing others. My point here is that as long as the tendency to naturalize social difference remains in place – as it does, at least implicitly, in so many areas of political and public discourse today – the categorical critique remains important. If powerful people think and act, and if institutions operate, as if races are real and racial inequalities are natural, then the argument that race differences are socially and politically signified, and that practising and perpetuating racial inequalities is unacceptable, is the appropriate anti-racist response. This response may no longer draw on simple binary formulations, but it is an approach that has to work with (even as it argues against) the mutually exclusive groupings – the categories – produced through the naturalization, racialization and unequalization of social difference.

The final part of the discussion goes on to consider the implications for this social constructionist, and anti-racist, thesis of a growing recognition that the dualism which inspired it – the nature/culture dichotomy – is itself collapsing in most key areas of social thought. The anti-racist project has depended on that dualism being in place. It has indeed helped reproduce that stark divide. The question I wish to raise, then, is this: if the nature/culture binary disappears, what defence does it leave us against the charge that the world's populations really are divided by nature – by physiology, anatomy, biology, or genetics – in ways which are significant for politics, incomes and wealth?

Colour-coding: the nature of society

There is much speculation on where, why and how the racialization of human populations originated, but little doubt that the colonialisms and imperialisms which underpinned European economic expansion in the nineteenth century are critically implicated. So too is a practice often located at the heart of geography – mapping. Maps identified vast areas of '*terra nullius*' (land belonging to nobody) as available for European powers to claim. Eurocentric maps were also used to label those who occupied such lands as different (how else could these first peoples be so comprehensively disenfranchised?). Such people were branded as uncivilized, exotic and, above all, colourful – red, yellow, brown, black – by cartographies that distanced 'white' Europeans from 'other' races, and from the atrocities exacted in return for European 'progress'. It soon became well established in philosophical and geographical, and in popular and political, thought that where people come from provides a useful index (is, indeed, an environmental determinant) of who they are, that who they are is inscribed on the body in a cluster of features indexed by skin pigment, and that this (together with clusters of other phenotypical signifiers) therefore says something fundamental about where people should (and should not) be. Livingston (1992, 1994) has charted this process in some detail. It is no exaggeration to suggest that the entire European colonial project was justified by appeal to the notion that places are different (in ways which impact on the nature of the people who occupy them), that races are real, and that the inequalities dividing these racialized groups are natural and inevitable.

This appeal to nature as an explanation for the inequalities dividing human populations was rooted in a growing intellectual excitement about nature, and especially about the spatial order of nature, during the nineteenth century. The enterprise of the day was to discover things, group them and rank them according to their position on the evolutionary scale. Elaborate hierarchical plant and animal classifications were developed which, as Kay Anderson (2000) has shown, soon raised the critical question of how, among sentient creatures – who are conscious and feel pain – some justification could be developed for placing *homo sapiens* above the rest. Amid much hand-wringing, the answer was generally held to rest in the capacity to reason, the ability to domesticate and the tendency to civilize. All these features enabled humans to distance themselves from, and place themselves above, the nature they were a part of, and to control it: to change and adapt this nature in ways that might improve the world.

This is what racialized the culture/nature debate. Civility and domesticity are not characteristics that can be measured and monitored in the same scientific way as lung capacity, organic complexity or mating behaviour. Anderson (1997) tells how the illusion that cultural values could be 'scientifically' objectified led to a number of problematic oppositions, not just between what was civilized and what was wild, what was domesticated and what was untamed, but also between the purity of the mind and the animality of the body, between citizen and savage, between education and ignorance, between morality and unruliness. The end point of this series of elisions – of this gradual conflation of biology with morality – was that some people were placed in lower positions than others on the scale of humanity, and some places were defined as less progressive than others on the scale of civility. This scale was, as Kobayashi (2003) has argued, powerfully spatialized in the influential geographies of Immanuel Kant which helped legitimize the grading and ranking of colour by latitude. Racial essentialism, then, was built on a naturalized geographical distinction between all that is civilized and cultured – the pinnacle of humanity, epitomized in 'white' European men – and all that is savage and wild, namely creatures (animals and peoples) and objects that were placed much further down, and much further away. At this 'lower', more distant, end of the scale some racialized groups were labelled as closer to animals than to humans, and some were dehumanized altogether, finding themselves effectively on the 'wrong' side of a newly emerging culture/nature divide.

So the culture/nature divide was not simply practised as a divide between minerals, plants and animals on the one hand and humans on the other. It was a divide which cut through human populations, on the pretext that some people simply had not had the opportunity (and possibly did not contain the capacity) to become 'civilized'. It was indeed a divide which aligned what it was to *be* human with what 'Western' societies had 'achieved' (Anderson 2001). This, of course, was highly convenient to the colonial effort. Just as humans domesticated nature in order to improve it, so colonial powers justified the appropriation of colonial lands in order to 'improve' them. The process was not one that appealed only to science for its legitimacy; its assumptions were embedded in the whole interlocking system of representations making up the cultural infrastructure of the imperial world. Said's (1993) essays on culture and imperialism trace this out in some detail. He shows, for example, how the Western novel and the colonial enterprise developed hand in hand: one was about the mapping and acquisition of physical spaces; the other about the appropriation of meaning and the control of creative imaginations. So even in the novels of Jane Austen, who wrote about the minutiae of domestic England, there are allusions to the 'facts' of Empire and assumptions about what this means. In *Persuasion*, for example, Austen describes Admiral Croft, who had been stationed for several years in the East Indies, as 'a native of Somersetshire, who, *having acquired a very handsome fortune*, was wishing to settle in his own country'. Captain Wentworth, on the other hand 'had no fortune … [but] he knew that he should soon have a ship and soon be on a station that would *lead to everything he wanted*' [emphasis added]. The point Said makes in reflecting on this is that whether writing about domestic niceties or overseas adventures, novels, like politics, were part of an interlocking system of racist practices and ideas which depicted the colonies as 'our' possessions, located in far-flung places, peopled by primitive, wild, exotic, eccentric or dangerous (not-quite-)human beings, and serving as a source of income.

What this history also shows is that the categorization of human populations into discrete racial groupings was part and parcel of the setting up of one of the most powerful binaries in Enlightenment thought: the distinction between nature and culture. The idea of race was only possible because of the way nature and culture were distinguished: because the base instincts of animality were assigned to a different and inferior realm than the moral order of humanity (see Anderson 2002). Thinking in 'black and white' was part of this, because these markings were made to represent the two extremes of a notional 'racial' hierarchy. But this black/white binary also informed a rather different historical moment: it was not only about naturalizing race difference; it was also part of a strategy for resisting racism. It was a vehicle for challenging the nature of, and in, race.

Black and white

As the twentieth century unfolded, sooner or later (and, above all, unevenly) social science mounted a powerful critique of the colonial project. Part of this process involved unpicking the elaborate theories of racial superiority and inferiority which claims to a natural order had legitimized. Social scientists attacked the appeal to nature that justified colonial exploitation, and to this end they sought to make a clear distinction between what does, and what does not, arise 'naturally'. The drive was on both to distinguish culture from nature, and to place all human affairs firmly in the former category.

This denaturing of human life was part of a much broader trajectory within social science. The nature/culture divide was employed strategically – in order to place responsibility for human affairs unequivocally in the hands of people themselves – in many areas of research, including the overlapping arenas of social and radical geography, in the 1960s and 1970s. Urban geography, for example, reacted against the legacy of an ecological account of city structure which invoked Darwinian ideas to 'naturalize' the links between immigration and segregation in the early part of the twentieth century. The effect of using terms like 'invasion', 'succession' and 'dispersal' to account for who lived where, not only helped reproduce the idea that race differences are real, and that difference causes problems, but also deflected attention away from the role of human agency in the process of city growth. This process of naturalization fitted neatly with the prevailing non-interventionist political economy, but it cast no light at all on the processes by which richer and poorer people, 'white' and 'black' residents, fared consistently and systematically better and worse in the struggle for resources and life chances in an unequal world. It was in an attempt to account for, and address, these inequalities that urban sociology, urban geography and urban theory banished ideas about nature from the discourse of the city, and were swept up instead by a radical edge of intellectual debate rooted very firmly in the world of capital, labour, politics and economics.

Likewise, in challenging the racisms that divided populations according to skin colour, and which sustained marked inequalities between the groups so racialized, most mainstream social research in the latter half of the twentieth century agreed – tacitly or explicitly – that studies of race and racism should draw a clear distinction between nature and society. This distinction was drawn precisely to make the point that any social, spatial, economic or political differentiation within the human populations of the world should be treated fundamentally and exclusively as a product of social, political and economic processes. This line of thinking therefore cemented the separation of nature from culture by repositioning the boundary between the two to ensure that all human beings were located firmly on the social side. The anti-racist case rested, in short, on all people being categorized as humans. What we now call the social constructionist approach to studying race and racism has its rationale in the argument that there is no natural explanation for the 'facts' of racial difference.

Miles' (1982) work on racism and migrant labour has probably put the case most fully and most forcefully. He makes the case that there are no boundaries in the human genome that correspond with the racial differences that so starkly divide the world. His argument, which has become the point of departure for much anti-racist social science, is that society is not naturally divided into racial groups, it is actively racialized by social, economic and political power relations. What this means is that appeals to nature have no place in our understanding of racial(ized) difference. If we want to understand how inequalities are racialized, how racial difference is made and reproduced, it is to sociology and politics, not biology and genetics, that we should turn.

This 'turning' changed the focus of enquiry from races to racisms; it changed the focus of social geography from 'how is race mapped onto space' to 'how does the economic, political and social organization of space produce and reproduce the idea of race, and the inequalities between racialized groups?' And it is by tackling this second kind of question that the binary black/white came to achieve its prominence in the research literature. As colonialism gave way to post-colonialism – in the mixing and annihilations of populations produced first as 'new worlds' were 'settled', and then through labour migrations from the periphery to the still-expanding economic core – the major inequality, the key economic and political distinction, was between 'white' Europeans and the rest. And for some time, for strategic reasons, 'the rest' became mainly, though not exclusively 'black'.

For more than half the twentieth century, for example, the label 'black' was legalized as a marker for exclusion and oppression in the *de jure* racial segregations enforced through the Jim Crow laws of the Southern States of the USA, and through Apartheid in South Africa. Challenging these laws meant working with these categories. Similarly, once (and in some places in order to secure the fact that) racial discrimination became illegal, the label 'black' was retained as a way of documenting and challenging the *de facto* exclusion of people of colour from full participation in the institutions and opportunities of 'white'-dominated societies. And the category 'black' has been used strategically by excluded groups as a focus for political strategies of resistance to racisms, for example in the development of Pan-Africanism, in the American Civil Rights movement, and in the deployment of 'black' in some parts of the British anti-racist movement. In all these areas working with the black/white binary has been part and parcel of a commitment to anti-racist principles.

What this brief overview suggests is that as the idea of race difference was wrested from the domain of nature, so the 'colourful' discourse of 'real' racial difference was replaced by an interest in forms of 'white' racism and strategies of 'black' resistance. Racial inequality was recognized to be a product of the uneven distribution of political and economic power, and a key axis around which this power struggle revolved was that dividing 'black' from 'white' – around a colour coding used to index the difference between those occupying a colonized periphery from those located in the space and spirit of a colonizing core. So the binary formulation black/white was not simply a problematic relic of Enlightenment thinking but was rather an integral part of a commitment to anti-racism – a step in the direction of civil and human rights, which stayed on the agenda well into the 1990s (Sharma 1996).

Working in 'black and white', even as an anti-racist strategy does (and should), nevertheless attract two key kinds of criticism. The first is its insensitivity to power struggles around racialized categories that do not fit into (and indeed have been actively excluded from) the categories black and white. Thinking in black and white excludes consideration of the historically situated oppositions around Jews/'Aryans', Western/Oriental and Christians/Muslims. It also glosses over a range of strategies of erasure practised against indigenous, native and first peoples in many parts of the colonized world. The reductionism inherent in the opposition black/white is further challenged also in the work and experiences of South Asians in Britain, whose positions and identities were made invisible by the black/white binary (see Modood 1994). One result of these debates is a growing recognition that just as racisms are diverse and changing, dividing populations in increasingly subtle yet often devastating ways, so the landscapes of racialization, and the positioning of 'strategic essentialisms', have become increasingly flexible and fine-grained. That is, where race categories are seized on to mobilize resistance against the racisms that created them, they too are responsive to context, and they too work in a spectrum between the poles of black and white. A signifier like 'Asian' – or subdivisions of this wider grouping – may therefore be needed, as Sharma (1996: 34) puts it, as 'one of many temporary positionalities that offer us strategic places from which to speak in this racist Britain'.

The key point here is that social and cultural, political and economic processes still discriminate, and that populations remain divided and damaged by racisms. As long as this remains the case, thinking and working with (albeit against) categories – even if they shift and change in time and space for a variety of strategic reasons – will remain high on the anti-racist agenda.

Something akin to the black/white binary – the use of categorical thinking to challenge essentialist practices – remains in place for the very best of ethical and moral reasons.

That being the case, a second criticism of black/white working requires attention, namely that there has been more scrutiny of the former than of the latter category. Recently this has reflected the growing presence, within a still-exclusionary academy, of people of colour who have adopted a range of speaking positions in order to articulate the diversity of identities and trajectories previously bundled up within the label 'black'. But equally, it expresses a longer-standing preoccupation among predominantly 'white' academics with documenting the worlds of people oppressed by racisms (people of colour, 'black' people, 'non-white' subjects) in scholarship which has, as Bonnett (2000) shows, had the effect (albeit unintentional) of 'normalizing' the equally racialized position of 'white' dominance. Bonnett's own project has been to expose the history and geography of the construction of whiteness, rendering its 'normality' problematic and so laying its dominance open to challenge and change. At the same time, his argument has been that 'despite the almost universal abandonment of explicit doctrines of white supremacy, and the adoption of anti-racist rhetoric as the lexicon of legitimacy by institutions the world over, whiteness continues to be reified as a racial and cultural norm' (Bonnett and Nayak 2002: 309). While this 'white racial norm' retains its capacity to adapt and survive, it is hard to escape the logic of thinking and acting with it (if only to challenge it), even though the motivation is ultimately to transcend it.

There are, nevertheless, some criticisms of constructionist critiques of race, no matter how fine-grained the categories on hand might be. For the most part these are about refining rather than rejecting this approach. This is true, for example, of the charge that constructionism has emphasized the social over the political, and that it has sometimes failed to link the material and cultural aspects of racial differentiation (see Kobayashi 2003). However, a more fundamental concern, which I have raised before (Smith 1999), and which is not circumvented by multiplying the number of racialized positionings we recognize, by challenging the normality and homogeneity of 'white', or by bringing a more material, political edge into the debate, is that constructionism necessarily remains trapped within the categories it seeks to challenge. It depends on drawing the distinction between nature and culture in order to argue *against* the process of social categorization (racialization) but it does not contain a prescription *for* a particular alternative. This need not be a problem while the boundary between culture and nature remains in place. With this as the bottom line, it is always possible to argue that divisions among humans are not natural but made. Within this logic, social inequality can never be regarded as the inevitable consequence of 'essential' distinctions. It is clear what anti-racism opposes. But as I go on to show, nature has now returned to society in some powerful and persuasive ways, and this is unsettling for the 'race' debate. People, nature, things and technologies are increasingly mixed up, even merged. This exposes some gaps in what once seemed a much simpler anti-racist logic. At the very least, this directs more attention to the question of what a world without racism could, or should, be like. At best, it requires some firming up of what anti-racism is *for*.

Social nature: de-racializing the landscape?

To recap: the culture/nature dualism was deeply inscribed in Enlightenment thinking and it was actively reproduced through much of the twentieth century, not least as a means of placing the responsibility for divisions and inequalities among peoples squarely in the realm of political, social and economic practices. But this way of thinking has been challenged in recent years as the culture/nature dualism has come unstuck, especially in the hands of geographers, whose discipline itself bridges the natural and social worlds. The literature is too big to tackle here, but the upshot is that the argument that nature is natural, set apart from the social, has been displaced by the return of holistic and organicist notions of how the world works. James Lovelock's Gaia and Arne Naess' deep ecology are part of this movement, as are Sarah Whatmore's (1999, 2002) hybrid geographies. The turn to nature has, moreover, engaged the critical edge of human geography. For example, a new dialogue has opened up between Marxism and environmentalism. This conversation has been propelled by the linkages between social and environmental justice forged by Harvey (1996), into a world where the natural and the social are indivisible (Castree 1995). In all these respects, the collapse of the culture/nature dualism seems intellectually well founded, somewhat overdue and entirely welcome. But it does raise some problems for constructionist theories of race, and therefore for the anti-racist project and the human rights that hinge on it.

As I have already argued, the separation of nature from culture is precisely what has been relied on in order to challenge the essentialisms that underpin powerful racial ideologies. Anti-racist social science has founded itself on the premiss that there is a world of nature which does not include people and that there is a world of people whose differences and inequalities are anything but naturally inspired. If this premiss is rescinded where does that leave the anti-racist argument? What happens to our ideas about race if we dare to let nature back in?

Daring to imagine a natural/natured social world is difficult in light of the legacy of xenophobia and genocide left by theories of natural social difference. The entire edifice of slavery, the 'evacuation' of Jews from the Aryan race, the wider eugenics movement, the practice of Apartheid, the dispossession and expulsion of Asians from Uganda, all this – and much, much more – has been justified by the idea that races are real; that racial differentiation is natural and inevitable, and that racial mixing is an undesirable trend. In light of this, it is entirely understandable that research and writings on the topics of race and racism have been last in the queue to take on board the collapse of the culture/nature dualism. Yet in a well-intentioned, politically critical bid to remove allusions to nature from our analytical vocabulary, we seem to have overlooked the fact that some of those who speak as, and for, oppressed peoples are using nature as a platform, sometimes bringing ideas from nature into political debates about the shape the future might take. What I want to suggest is that by engaging with these debates there is a case for bringing nature back into society in a way that advances rather than diminishes the anti-racist case.

In exploring this possibility, I am mindful of bell hooks' concern when, identifying herself with all peoples oppressed by colonialism, imperialism and essentialism, she makes the point '*we* must determine who we will be'. I am therefore wary – as a (now) middle class academic, concerned with what it has been to be 'white' – of appearing to speak for (or worse still, instead of) colleagues, or folk, whose history has been otherwise. However, as I argued above, while the advent of 'white studies' is welcome and has been important, it is still an approach that, even as it exposes the heterogeneity beneath the surface of 'black' and 'white', is bound into the categorical thinking these divisions imply. So it is important, I think, also to engage in the conversation sparked off by those voices of self-determination to which bell hooks refers. To this end I want to reflect on the (re)integration of nature with culture which has been effected by artists of colour whose work may offer – among very many other things – some ways of imagining the world that do not depend on the opposition between black and white, some visions of a future that does not depend on the racialization and unequalization of difference. I am not suggesting that these artists deliberately set out to do this, and I am sure that their work could be interpreted in many other ways. I do not, then, claim to be 'right'; but I do claim the right to imagine.

There is little space here to elaborate on why art might matter for social science (though see Smith 1997). Film-maker Spike Lee puts it quite directly:

> Presently in America a war is being fought ... It's a war on the battleground of culture ... At stake is the way to control the way people think or not think, act or be passive ... In this war, it's gonna be down to the artist. (Lee 1993: Preface)

bell hooks (1995) makes the point in a more analytical way, showing how the link between art and power can be both oppressive and emancipatory. On the one hand, she says, 'art expresses the horrors of domination when they are too painful to speak of', so that 'if one could make a people lose ... their power to make art, then the work of subjugation ... [would be] complete'. But on the other hand she points out that 'the creation and public sharing of art is essential to any practice of freedom', to which end 'aesthetics nurture the spirit ... healing wounds inflicted by ... imperialist, racist and sexist domination'.

So artworks are important in understanding and managing the power relations which structure lives and reproduce inequality. And where formal power is withheld, artworks are a particularly powerful means of political expression. Artworks then can give voice to otherwise silenced positions; can begin to will alternative futures into being. To explore this I shall take two of many possible examples to suggest how, by opening up some routes between nature and culture, we might simultaneously find ways of working between, or without, the race categories that have been so ingrained in the ways of the world.

The dualistic thinking which split nature off from culture (which for a period was helpful to the anti-racist movement) also deflected attention from two other key themes: the denigration and devaluation of non-human animals (which has been overlooked by what has sometimes been termed the 'humanocentrism' of anti-racism), and the power of the body. By challenging the former and harnessing the latter, the artworks discussed below might offer a way to recast the terms of, and ultimately move beyond, the binary and categorical thinking that have ensnared the race debate.

Replacing nature

In most European and North American societies people of colour live disproportionately in cities, often in highly segregated communities. This is a fact, and an idea, which some artists have been keen to challenge, questioning the received wisdoms that render vast swathes of landscape 'white'. Kinsman (1995), for example, draws attention to the work of the British photographer Ingrid Pollard, whose 'Pastoral Interludes' both document and question the opposition between 'white rural' and 'black inner city' environments (see also Smith 1999). On the other hand, while researchers have been busy distancing urban studies from ecological images, artists who identify with oppressed, disadvantaged and racialized urban communities have readily seized on natural metaphors and biological allusions in setting out their experience of urban life. 'The message', written and performed as a distinctively 'black' music in 1982 by rap artists Grandmaster Flash and the Furious Five, is an example of this.

> 'It's like a jungle sometimes,
> it makes me wonder
> how I keep from going under ...
>
> broken glass everywhere,
> people pissing on the stairs
> you know they just don't care ...
> I can't take the smell, can't take the noise,
> I got no money to move on, I guess I got no choice

This same metaphor of the urban jungle is taken up by Spike Lee in a series of films which similarly draw on biological allusion in order to speak about urban life today. I hesitated to take Lee's work as a case study here because some critics have argued that his films trivialize racism, or at least shy away from documenting the complexity of its forms, mechanisms and effects (see Diawara 1993; Gilroy 1993; Guerrero 1993; Reid 1997). Additionally, there is the charge that he fails to take women of colour seriously in his films. On the other hand, Lee's own view is that 'Young Black people are coming together, as one, to make our own films. It's never been done before on this scale. What we're doing is revolutionary and courageous' (Lee 1989: 63). Building from this, and thinking particularly about *Do the Right Thing* (1989) and to a lesser extent *Jungle Fever* (1991), there is one set of ideas which I do want to engage with, because they might raise the possibility of creating a new kind of space for the idea of race.

The starting point is set by Lee in a speech to the Imagination Conference in San Francisco in 1996, which outlines his aims as a film-maker:

> Growing up in this country, the rich culture I saw in my neighborhood, in my family – I didn't see that on television or on the movie screen. It was always my ambition that if I was successful I would try to portray a truthful portrait of African Americans in this country … (www.inmotionmagazine.com/slee.html)

The 'truthful portrait' has to do with the dignity, legitimacy and positive presence of African-American life in North America. In crafting this portrait, it seems to me that Lee's ideas could help us imagine a society in which race categories are made to lose their power to order life chances and opportunities.

At first glance, this seems unlikely because Lee so often confronts the stark inevitability of racialized difference, especially in *Jungle Fever*, which is about the doomed affair of a black man and a white woman, whose attraction may be based more on curiosity than love. Furthermore, Lee himself worked explicitly with the idea of black people's distinction and autonomy as he strove to establish an independent presence in the film industry. The sequence of racial slurs at the heart of *Do the Right Thing* also seems encased in this world. Mookie, the film's African-American lead (played by Spike Lee himself) launches a barrage ('Dago, wop, garlic-breath …') against Pino, son of Italian-American Sal, who owns the neighbourhood's only pizzeria. Pino has no sooner retorted ('You gold-teeth, gold-chain-wearing, fried-chicken-and-biscuit-eatin' …') when Stevie the Puerto Rican cuts in against the Korean and Chinese small business owners ('You slant-eyed, me-no-speak-American, own every fruit and vegetable stand …'), and is himself cut off by white police officer Long ('Goya-bean eating, fifteen in a car …'). Finally one of the Koreans swears at a Jew. This is a part of the film that could be criticized for reducing racism to prejudiced, individualized name-calling, downplaying its deep-seated institutional structures and consequences. However, what is thought-provoking about it in the context of this chapter is that it bursts out of an exchange between Mookie and Pino which hinges starkly around the difference between black and white. It sets the scene, albeit in a highly charged way, for disturbing a well-worn dualism in a neighbourhood which Lee himself wanted to 'have the feel of the different cultures that make up the city' (1989: 29).

As the film unfolds it seems possible that Lee's project is not only – or even – about reproduction of established racial stereotypes. Instead it sets out to challenge them, not by denying a history of difference or ignoring the racialization of inequality, but rather by making the film more about the dignification of what it means to occupy the spaces in which (in this case) predominantly black Americans live, than about a response to the denigration that created these spaces in the first place. In this, Lee is questioning the idea, ingrained in so much of popular culture, that African-Americans are either stars, in the fantasy world of the unraced elite, or 'just niggers' in the criminal underclass. In contrast, Lee's aim appears to be to portray the African-American, and other racialized, working classes as the American norm: 'Contrary to popular belief,' he writes 'we work'. The film, he said, will have 'no welfare rolls ... just hardworking people trying to make a living'. It was not to be a film about saints and demons, but rather

> [i]t was about Black underclass in Bed-Stuy, a community that has some of the highest unemployment, infant mortality, and drug-related homicide rates in New York City. We're talking about people who live in the bowels of the social-economic system, but still live with dignity and humor. (1989: 109)

The film is set in the midst of the 'urban jungle' on the hottest day of the year, yet by emphasizing the ordinary, it could be argued that Lee is blurring the boundaries between the social and natural worlds in a way that avoids – or at least challenges and redefines – the hierarchy from things civilized, domesticated, human and 'white', to things savage, wild, animalistic and 'black'. Lee's 'jungle' is characterized by civility and dignity; Lee's ghetto is not 'naturally' dark and different, it is a place where ordinary Americans spend much of their lives doing ordinary things despite the way society has treated them, and despite the way in which 'the ghetto' is popularly imagined. There is much more to the film than this, of course, including scenes of violence which might be interpreted as reproducing rather than breaking down key stereotypes. But it is the gesture towards the normalization and dignification of segregated spaces that I want to dwell on because this contains the bones of a strategy for reintegrating nature with culture in a way that challenges the kind of difference that race makes.

This, interestingly, is a challenge that constructionist approaches have not yet fully met. Constructionism has made the case for civil rights and equal opportunities, but none of this has yet removed the tenacity of race as a principle of social organization and an axis of inequality. This is what seems to underpin Gilroy's (1998: 843) concern that 'however essentially noble the ideal of anti-racism might be, it does not communicate anything affirmative'. What Lee's remixing of nature with culture suggests to me is that the social constructionist argument has not – in studies of race and racism at least – been taken far enough. The real achievement of this approach was to reposition the boundary between nature and culture so that all people were on the same side: differentiated by politics and economy, not by biology. But if we go on to recognize that ever having such a boundary, dividing people who are in control from nature which is controlled, is itself a product of history rather than a property of nature, then removing that boundary could pave the way to erasing the inequality across it. If 'nature' is not distinct from and subordinate to 'culture', then nothing and no one is eligible for denigration on the grounds of being closer to the former rather than to the latter.

The foundations of a theoretical case for this approach are laid by Wolch et al. (1995) in their advocacy of a trans-species urban theory, proposing an environmental ethic that recognizes the fundamental linkages between human justice and justice for animals. Philo (1995) similarly argues that there is a parallel between how we depict and react to animals in the city and how we depict and react to a range of marginalized *social* groupings; and Emel (1995) points out that sexism, racism and animal abuse are inter-related. What all these authors pick up on is the way that so much of the violence against essentialized groups who are (defined as) other-than-white, and women, has historically been legitimized by investing them with less-than-human qualities (by associating them, for example, with wild, untamed, animalistic tendencies) and by assuming that nature – identified as the non-human living world – is there for the satisfaction, gratification and service of 'Man'. This raises a question, which is powerfully addressed by Kay Anderson (1995) in her seminal paper on the Adelaide Zoo, of why the boundary between nature and culture was put there in the first place. If we take it away and look critically at how people treat, use and abuse animals and environments, we might better understand the exploitations and oppressions that we find in the human world. Perhaps the way forward is not so much to emphasize the difference, as to look for the similarities, between people and non-human animals. The implication here is that if we can give all sentient beings a more dignified place in the world, then the world will also be a more dignified place for people to occupy.

If living things that are not human were not seen as inferior – but just as different – then there would be no grounds on which to use the human/non-human gradient to justify subjugation and oppression. If we take away the notion that violent and harmful treatment is justifiable as long as the targets are not human, then the idea of 'dehumanization' loses its cutting edge; certainly there can be no claim to legitimacy for actions and behaviours that try to assert the superiority of some people over others. And this in the end is one part of Spike Lee's conversation: his urban jungle is a space in which the opposition between the civilized city and the racialized ghetto dissolves, along with the debate over constructionism and essentialism as the basis for racial categorizations and identifications. Once the dualisms savage/civilized, wild/domesticated, bestial/human, black/white are collapsed, then whether people regard themselves or others as 'raced' or not takes on a new meaning. This meaning need not deny people's sense of identity, cohesion, purpose or destiny; it need not attract, as the notion of 'hybridity' has, the charge that just as excluded groups are seizing on the idea of race for strategic purposes, powerful voices are denying that such categories exist (cf. Bonnett and Nayak 2002; Smith 1999). The point is, that in this new 'natural' order, as far as resource allocation and life chances are concerned, skin colour, facial appearance, shared religions, language or ancestry have the potential to become no more relevant as a principle of inclusion or exclusion than shoe size, handedness or fingerprints.

Remembering the body

Some readings/watchings of Spike Lee's works may help us imagine the possibility of a world in which race categories lose their potency as a marker for inequality. Now I want to consider an artist whose vision could nudge us one step further, in a direction uncomfortably signposted by Gilroy (1998) in his enigmatic essay 'Race ends here'. This journey starts with the recognition that it is not just non-human animal life that was written out of social theory by the nature/culture dualism. The human body was also effaced. As 'race' is above all a marking that is written on the body, the character of the body is a key consideration for any reworking of this set of ideas. A powerful testimony to this can be envisioned in the paintings of Jean-Michel Basquiat whose work might inspire those who can see, to think beyond the markings of race.

Basquiat was born in 1960. His father is Haitian, his mother Puerto Rican. His best-known paintings were made in the early 1980s on the subject of New York. He began as a graffiti artist, and most of his completed work is rooted in his experience of the street and his life in that city. Yet his preoccupation is not with the street or the 'hood, but with the body. For the purposes of my argument in this chapter, a key interpretation of his work is that it begins to imagine ways of representing the body as something other than a signifier for race difference (see Mirzoff 1995).

Although this discussion is primarily about Basquiat's attempts to unmark or de-racialize the body, it needs to be set in context. Basquiat's starting point is very much with the oppressions and inequalities of urban life for people of colour in the USA, and for 'black' America in particular. His paintings often focus on half-formed, stick-like black bodies (sometimes himself), which may symbolize the incompleteness or destruction of the psyche and identities of those who experience themselves as 'black', or as part of some other grouping which that label has come to represent. And in paintings like 'The Irony of a Negro Policeman' (1981) Basquiat vividly illustrates the tragedy of black compliance, flagging the processes through which oppression reproduces itself – here, because of the way that racism is institutionalized within the criminal justice system in ways that are at least partly beyond the control of the individuals that operate that system.

However, a different idea may be threading its way through some of Basquiat's other works. The seeds are sown in his parodies of Manet's iconic painting of Olympia (1865). This (Manet's work) is a painting of a beautiful seductive woman reclining unclothed, and attended by a black woman servant. This vision epitomizes the pre-occupation of Western art with images of beauty – an idea personified in, and embodied by, youthful, sensual, white, female, figures.

But Basquiat's painting 'Three Quarters of Olympia Minus the Servant' (1982) exposes this pre-occupation as a racist illusion. Figure 6.1 is a painting of Olympia alone.

FIGURE 6.1 *Three Quarters of Olympia Minus the Servant*, 1982 (Jean-Michel Basquiat). DACS, London 2004

Without the servant, she is ugly, disfigured and largely disembodied. It seems that the perfection of the body in Western culture is only sustainable – is only imaginable – in opposition to the 'imperfect' racialized body of the exotic but servile Other. The crushed features of Basquiat's Olympia signal the extent to which Manet's work takes the presence of the maid for granted. This 'taking for granted' allows the viewer to ignore the processes of colonialism that made the juxtaposition of complete white beauty against servile black attendant possible in the first place. In contrast, the reality of this history, which is laid out in Basquiat's painting, is that Olympia (and the tradition she represents) is also diminished by her compliance with the oppositions that colonialism, racism and other essentialist oppressions set up. Her body is brutalized and dismembered too (and is no longer of interest to the male gaze for which it was designed). Her expression is twisted and pained, as if, now the maid has gone, she has learned to see the full horror of her position.

In a second painting (Figure 6.2), which is untitled but bears the caption 'Detail of Maid from Olympia' (1982), only the maid features, and she is central to the canvas. For me, the maid is now a much more dignified figure than Olympia, with a grace and integrity that Olympia will always be denied.

FIGURE 6.2 *Untitled (Detail of Maid from Olympia)*, 1982 (Jean-Michel Basquiat). DACS, London 2004

Prising her apart from Olympia has not, however, made her whole. hooks (1995) describes her as 'ugly and grotesque', framed in a blaze of red paint which 'drips like blood' and embodying still the anguish inflicted by the colonization of the black body and mind.

In Manet's work the 'perfect' body (white and Western) has only been made imaginable in opposition to the 'imperfect' (black, colonized) other. Basquiat has taken away the imagined other, and brutalized the once-perfect self. While this is undoubtedly very much about some 'old' oppositions, it raises too the possibility of forging a new kind of connection between Olympia and the Maid (who is no longer a servant). There is some possibility here that Basquiat is mixing up – perhaps beginning to take apart – the familiar race codings of the visual order.

In some other paintings Basquiat seems to take this further. Scattered among the black stick-like figures for which he is so well known are two other kinds of image. Untitled works painted in 1981 and 1982 include a variety of multi-coloured figures. One is a head, painted in black, yellow, red, orange and blue with a pink right eyelid. Another is a whole person in yellow, green, red, black and brown. These colours undoubtedly build on layers of black history. Black, red and green, for example, are the colours of the United Negro Improvement Association, yellow, green and red are the colours of the Ethiopian flag. But multiple-colourings around this starting point speak too of a rethinking of the colour-coding that has dogged so much of human history. That is, he moves beyond black and white, and, crucially beyond a single colour for a single body. People's markings are all mixed up, perhaps giving shape and form to Bhabha's (1994) hopes for a more open, hybrid culture: a way of being that does not rely on racialized oppositions.

When Gilroy (1998: 840) dreams that 'the time of "race" may be coming to a close', he hints at the unmarking and revisioning signalled by new technologies, especially innovations in multidimensional body imaging, which 'impact on the ways that embodied humanity is imagined and upon the status of bio-racial differences that vanish at these levels of resolution' (p. 846). Without the aid of these devices, it seems to me that Basquiat is pursuing much the same ends. A set of untitled figures painted in 1981 (a red man, a black man and a figure in both these colours) all provide a glimpse below the surface of the body at bones, organs and sinews – aspects of the body which signal points of similarity rather than difference among humans. At times, Basquiat even seems to strip the flesh away from the body as if it might be possible to start again. 'Riding with Death' (1988) (Figure 6.3) is the starkest example of this: a skeletal figure in full flight on a skeletal horse. It contains allusions to the classic symbols of death and destruction, but equally it resonates with classic tales of renewal and transformation. It is paintings like this that prompt Mirzoff (1995) to suggest that Basquiat comes closer, with his paint, brush and bodily representations, than any written philosophical or intellectual text to the possibility of imagining an alternative to the oppressive race codings of the past. Working

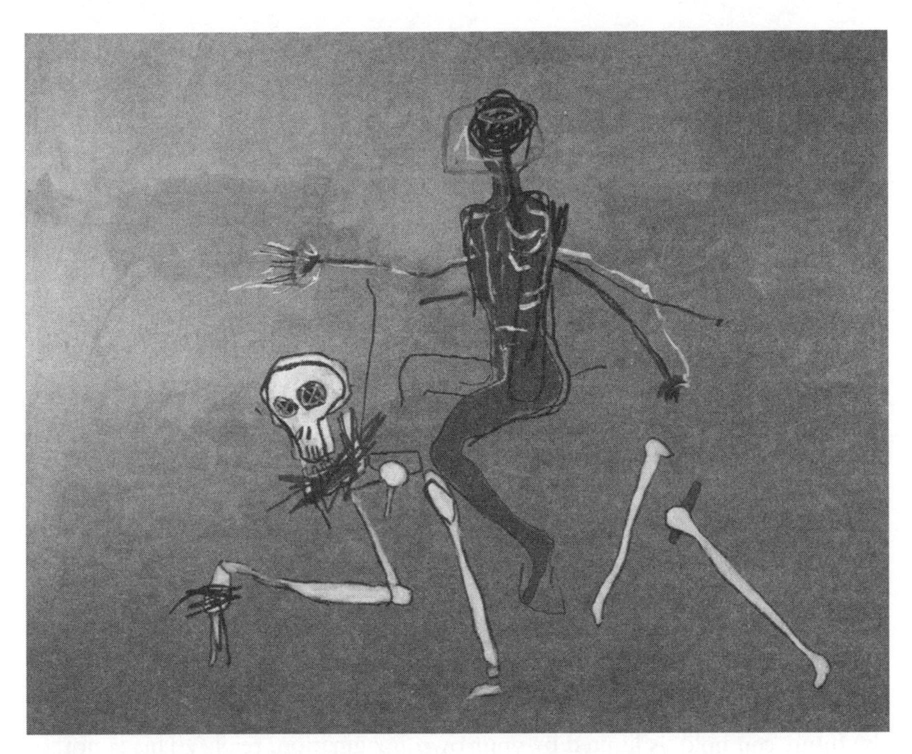

FIGURE 6.3 *Riding with Death*, 1988 (Jean-Michel Basquiat). DACS, London 2004

through the body, Basquiat brings an element of nature – the body itself – into society in a way that challenges rather than legitimizes the old social order.

One interpretation of some of Basquiat's work is that re-integrating nature with culture in the conceptualization of race has emancipatory rather than oppressive potential. That is, if we could imagine a body unmarked by race, perhaps we could also imagine (as some novelists have) a post-colonial map, a non-segregated city, and, in the end, a geography where people are not limited in where they can live, how often or how far they can move, how they work and who they work for, what they can earn, what they are entitled to, and who they marry, eat with and play with, by the markings on their bodies.

Concluding remarks

In the end there is a powerful argument for abolishing the binary black/white and for changing the meaning, and perhaps altogether dissolving, the racial categorizations that are indexed by this distinction. This, however, is a cautious conclusion, because it is based on an imaginary world. The 'real' world is still divided by racism, and race categories are still taken-for-granted in all walks of life, as are the inequalities they help legitimize. So a holding position is needed, and this must surely continue to be organized around some kind of constructionist and/or materialist critique of how differences are recognized, made and unequalized, as well as around some kinds of strategic essentialism. One way or another, the categorical oppositions through which racism divides societies need to be challenged in their own terms.

On the other hand, the capacity to imagine and perform something differently is a critical part of changing the status quo. In the late nineteenth century it would have been hard to believe that 'races' might be socially and politically constructed, yet for most of the twentieth century, we took that for granted and used it as a platform on which to resist racism. In the mid-twentieth century, it would have been hard to imagine referring to nature in the context of social, urban and political studies, but today it is unusual *not* to find 'nature' as a key word in most mainstream journals. As the twenty-first century gathers momentum, it is hard to conceive of a future without racisms, yet anti-racism has only ever been a means to an end. What it has lacked is a vision of what that end might be. The death of dualistic thinking, and especially the collapse of the culture/nature divide, suggests that this can and must change. What this chapter has tried to recognize is that, in speculating on this, artists can go where academics dare not. Speaking over one of his Modern Day Jazz Stories, saxophonist Courtney Pine argues that what the future can hold 'is limited by your own imagination, really'. This is not, of course, the only limit: but it is perhaps one whose tenacity should be challenged.

FURTHER READING

Anderson, K. (2001) 'The nature of "race"' in N. Castree and B. Braun (eds), *Social Nature: Theory, Practice and Politics.* Oxford: Blackwell. pp. 64–83

Bonnett, A. and Nayak, A. (2002) 'Cultural geographies of racialization – the territory of race', in K. Anderson, M. Domosh, S. Pile and N. Thrift (eds), *Handbook of Cultural Geography.* London: Sage. pp. 300–12

Gilroy, P. (1998) 'Race ends here', *Ethnic and Racial Studies* 21: 838–47

Mirzoff, N. (1995) *Bodyscape. Art, Modernity and the Ideal Figure.* London and New York: Routledge

Smith, S.J. (1999) 'Society–space', in P. Cloke, P. Crang and M. Goodwin (eds), *Introducing Human Geographies.* London: Arnold

Wolch, J., West, K. and Gaines, T.E. (1995) 'Transspecies urban theory', *Society and Space* 13: 735–69

References

Anderson, K. (1997) Culture and nature at the Adelaide Zoo: at the frontiers of 'human' geography. *Transactions of the Institute of British Geographers* NS 20: 275–94.

Anderson, K. (2000) The beast within: race, humanity and animality. *Environment and Planning D: Society and Space* 18: 301–20.

Anderson, K. (2001) The nature of 'race'. In N. Castree and B. Braun (eds), *Social Nature: Theory, Practice and Politics*. Oxford: Blackwell. pp. 64–83.

Anderson, K. (2002) The racialization of difference: enlarging the storyfield. *Professional Geographer* 54: 25–30.

Bhabha, H. (1994) *The Location of Culture*. London and New York: Routledge.

Bonnett, A. (2000) *White Identities. Historical and International Perspectives*. London: Prentice-Hall.

Bonnett, A. and Nayak, A. (2002) Cultural geographies of racialization – the territory of race. In K. Anderson, M. Domosh, S. Pile and N. Thrift (eds), *Handbook of Cultural Geography*. London: Sage. pp. 300–12.

Castree, Noel (1995) The nature of produced nature: materiality and knowledge construction in Marxism. *Antipode* 27: 12–48.

Diawara, M. (ed.) (1993) *Black American Cinema*. New York and London: Routledge.

Emel, J. (1995) Are you man enough, big and bad enough? Ecofeminism and wolf eradication in the USA. *Society and Space* 13: 707–34.

Gilroy, P. (1993) Spiking the argument. Spike Lee and the limits of racial community. In *Small Acts*. London: Serpent's Tail. pp. 183–91.

Gilroy, P. (1998) Race ends here. *Ethnic and Racial Studies* 21: 838–47.

Guerrero, E. (ed.) (1993) *Framing Blackness*. Philadelphia: Temple University Press.

Harvey, D. (1996) *Justice, Nature and the Politics of Difference*. Oxford: Blackwell.

hooks, b. (1995) *Art on My Mind. Visual Politics*. New York: The New Press. esp. Introduction: art matters, pp. xi–xvi, Art on my mind, pp. 1–9, Altars of sacrifice: re-membering Basquiat, pp. 35–48.

Kinsman, P. (1995) Landscape, race and national identity: the photography of Ingrid Pollard. *Area* 27: 300–10.

Kobayashi, A. (2003) The construction of geographical knowledge – racialization, spatialization. In K. Anderson, M. Domosh, S. Pile and N. Thrift (eds), *Handbook of Cultural Geography*. London: Sage. pp. 544–56.

Lee, S. (1989) *Do the Right Thing*. Companion volume to the Universal Pictures film. New York: Fireside.

Lee, S. with Wiley, P. (1993) *By Any Means Necessary*. Vintage: London.

Lee, S. (1996) Untitled talk delivered to the Imagination Conference, San Francisco, June 8 (www.inmotionmagazine.com/slee.html).

Livingston, D.N. (1992) *The Geographical Tradition*. Oxford: Blackwell.

Livingston, D.N. (1994) Tropical climate and moral hygiene: the anatomy of a Victorian debate. *British Journal for the History of Science* 32: 93–110.

Miles, R. (1982) *Racism and Migrant Labour*. London: Routledge and Kegan Paul.

Mirzoff, N. (1995) *Bodyscape. Art, Modernity and the Ideal Figure*. London and New York: Routledge.

Modood, T. (1994) Political blackness and British Asians. *Sociology* 28: 859–76.
Philo, C. (1995) Animals, geography and the city: notes on inclusion and exclusion. *Environment and Planning D: Society and Space* 13: 655–81.
Reid, M.A. (ed.) (1997) *Spike Lee's Do the Right Thing*. Cambridge: Cambridge University Press.
Said, E. (1993) *Culture and Imperialism*. London: Chatto and Windus.
Sharma, S., Hutnyk, J. and Sharma, A. (eds) (1996) *Dis-Orienting Rhythms. The Politics of the New Asian Dance Music*. London and New Jersey: Zed Books.
Sharma, S. (1996) Noisy Asians or 'Asian Noise'? In S. Sharma, J. Hutnyk and A. Sharma (eds), *Dis-Orienting Rhythms. The Politics of the New Asian Dance Music*. London and New Jersey: Zed Books. pp. 32–57.
Smith, S.J. (1997) Beyond geography's visible worlds: a cultural politics of music. *Progress in Human Geography* 21: 502–29.
Smith, S.J. (1999) Society-space. In P. Cloke, P. Crang and M. Goodwin (eds), *Introducing Human Geographies*. London: Arnold.
Whatmore, S. (1999) Hybrid geographies: rethinking the human in human geography. In D. Massey, J. Allen and P. Sarre (eds), *Human Geography Today*. Cambridge: Polity Press. pp. 22–39.
Whatmore, S. (2002) *Hybrid Geographies. Natures, Cultures, Spaces*. London: Sage.
Wolch, J., West, K. and Gaines, T.E. (1995) Transspecies urban theory. *Society and Space* 13: 735–69.

7 Man:Woman

Lynda Johnston

I want to start this chapter with a story about domestic cat and dog accessories. My dog's veterinarian clinic is selling soft toys – in the shape of dogs – for cats. The toy is called 'Doggy Dress 'N Drag' and comes with instructions: 'You dress him, your cat drags him'. The promotional packaging claims Doggy Dress 'N Drag to be the 'world's first accessorized cat toy' and features an image of a handbag. The dog comes 'dressed' in a skirt, which can be removed, and the packaging indicates that he is the same clothing size as Barbie. I selected a Doggy Dress 'N Drag called 'Big Yellow Bob' and I gave my friend 'Swanky Ken Dog' (also available is 'Frank', see www.kittyhoots.com). My dog, Big Yellow Bob, is discursively produced as masculine (by naming as well as markers of 'aggressiveness' such as bared teeth and a black eye). The dog is also feminized by its skirt which contains a repeated pattern of pink kitten heads. In short, Big Yellow Bob and Swanky Ken Dog seem a bit queer. Clearly, the makers of these cat toys are playing with notions of masculinity and femininity, heterosexuality and homosexuality. The play on gender and sexuality gains cultural currency when considered alongside animal 'norms' such as cat/dog rivalry. This toy makes fun of dogs because gnarly looking dogs like Big Yellow Bob and Swanky Ken Dog should not be feminized – made fun of – by the wearing of a skirt or, like Swanky Ken Dog, a bra. In other words, your cat can have a good old snigger – a kitty hoot – at the dog's expense.

Before I put away my animal accessories, I understand these toys to be queering notions of man/woman, heterosexuality/homosexuality. In this chapter I wish to follow Jon Binnie's (1997: 223) direction 'towards a queer epistemology' and adopt a camp appreciation for the ways in which notions of man/woman and homosexual/heterosexual might be produced and understood. The motivation for this approach is twofold. First, I am concerned with the ways in which the production of knowledge in geography rests on gendered and sexualized binaries and argue that certain forms of hegemonic masculinity and heterosexuality are operating to exclude and to 'Other' particular knowledges. I outline how man/woman binary thinking has been challenged by feminist geographers by drawing on work that explicitly genders and sexualizes knowledge.

Second, the performativity of bodies is under consideration here because of the disruptive potential to the binaries man/woman and hetero/homo. Like Binnie (1997: 228) I am dedicated to 'a greater commitment towards embodiment in our work'. In recent years geographers (see Longhurst 1997 for a review of this work) have become fascinated with embodiment and 'the body' has become a useful resource to articulate, for example, politics of difference. Within much of this work, 'the call to understand the body is often simultaneously a call for the fluidity of subjectivity, [and] for the instability of the binary of sexual difference' (Callard 1998: 387). Throughout this chapter I refer to the ways in which the man/woman binary continues to be crucial to the construction of geographical knowledge. At the same time, I point to the slippages, cracks and possibilities of troubling this binary. It is worth noting that I do not claim that one can easily and unproblematically step out of binaries such as man/woman, but I acknowledge that camp can productively work to undermine values and truths associated with gendered and sexualized bodies and spaces.

For a number of years I have been interested in gendered and sexualized bodies and spaces: for example, lesbians and domestic space (Johnston and Valentine 1995), female body builders (Johnston 1996, 1998), and gay pride parades (Johnston 1997, 2001, 2002). Much of this work examines the ways in which bodies and spaces are materially and discursively produced in particular places (and vice versa). Recently, I have turned to notions of camp because it enables me to continue to trouble that which is considered natural and to highlight relationships between power, knowledge, subjectivities and spaces. 'Camp contains an explicit commentary on feats of survival in a world dominated by the taste, interest and definitions of others' (Sontag 1983: 144).

I outline some of the debates that have contributed to the gendering of geography's dualisms, specifically how some topics have been associated with the mind, masculinity and heterosexuality and others have been associated with the body, femininity and homosexuality. The man/woman binary has its origins in early philosophies. I point to the cost of this binary and suggest that there have been several costs to geography as a consequence. I then explore the possibilities of a queer epistemology and attempt to reflect critically on some of my work involving gendered and sexualized bodies of gay pride parades. Throughout this chapter I adopt a camp appreciation of bodies and geography in order to trouble the binaries man/woman and heterosexual/homosexual.

Man/woman knowledge

Many geographers have questioned the ways in which the binary of man/woman is crucial to the construction of knowledge and discourse in geography (see, for example, Berg 1994; Bondi 1990, 1992; Bondi and

Domosh 1992; Johnson 1994; Johnston 1996; McDowell 1990, 1991; Longhurst 1995, 2001a; Rose 1993).This work stems from feminist theorists who have argued that the dualistic structure of Western thought is gendered as man/woman (Bordo 1986; Jay 1981; Lloyd 1993). Understanding the binary structured as gendered – and how it came to be gendered – enables a focus on the ways in which certain knowledge becomes aligned with masculinity (and privileged) and how other knowledge becomes aligned with femininity (and devalued). Here I outline some of the power relations that exist between Western scientific methods and hegemonic masculinist discourse by first tracing the history of gendered binaries.

Feminist philosopher Genevieve Lloyd (1993) examines the work of various philosophers (such as, Plato, Aristotle, Bacon, Philo, Augustine, Aquinas, Descartes, Hume, Rousseau, Kant, Hegel, Sartre and de Beauvoir) in order to highlight that since the inception of philosophy as a discipline in Greece, the maleness has been associated with reason, the mind and abstraction, and femaleness with irrationality, the body and materiality. The early Greeks associated women with reproduction and nature but men were understood to be able to transcend their bodily functions (nature) because of their capacity for reason. Plato was reported as saying that women imitate the earth (in Lloyd 1993). Lloyd traces this man/woman binary through to the sixth century BC when the Pythagoreans linked maleness to notions of enlightenment, orderliness and goodness. Femaleness was linked with darkness, disorderliness and badness. Explicit in this construction of opposites is the understanding that woman is to be avoided as she is associated with negative and 'dangerous' qualities, while man gains the highest status and is associated with virtuous qualities.

Geographer Larry Berg (1994: 249) offers a history of binary thinking and the ways in which it has become gendered. He refers to the Pythagorean Table of Opposites, which I include here (Figure 7.1) in order to illustrate positive and negative associations of Man/Woman.

Feminist Nancy Jay (1981) notes that dualistic thinking became embedded in Western philosophy when Aristotle (384–322 BC) outlined his laws of

male	–	female
limit	–	unlimit
light	–	dark
good	–	bad
one	–	many
right	–	left
rest	–	motion
straight	–	curved
odd	–	even

FIGURE 7.1 The Table of Opposites (Berg 1994: 249, adapted from Cavendish 1964 and Lloyd 1993)

formal logical thought. The most crucial rules are 'the Principle of Identity (if anything is A, then it is A); the Principle of Contradiction (nothing can be both A and not-A); and the Principle of the Excluded Middle (anything, and everything, must be either A or not-A)' (Jay 1981: 42).

These opposing positions gained more credibility during the Enlightenment and Descartes' theory of separation between the knower and the known. Susan Bordo (1986) calls this the 'cartesian masculinisation of thought'. Bordo (1986: 450) argues that this separation 'made possible the complete intellectual transcendence of the body'. 'Masculinist rationality is a form of knowledge which assumes a knower who believes he can separate himself from his body, emotions, values, past experiences, and so on' (Longhurst 1995: 98). Berg (1994: 249) adds that 'the mind was already associated with the masculine and the body was associated with the feminine, thus Descartes' mind/body dualism laid the conceptual groundwork for the masculine rational transcendence of the feminine irrational'.

According to feminist philosopher Elizabeth Grosz (1989), dualisms are part of a continuous spectrum that has been divided into self-contained elements which exist in opposition to each other. Grosz (1989: vxi) suggests that: 'when the system of boundaries or divisions operates by means of the construction of binaries or pairs of opposed terms, these terms are not only mutually exclusive, but also mutually exhaustive'. It is important to highlight here that the two sides of the dualism are not unrelated. If one side is represented by 'A', then its opposite will not be something from a different set of category relationships, say 'B', but rather will be a conceptualization of what 'A' is not, say 'A-'. The sides of the dualism, therefore, have an epistemological relation. This is a mode of knowing in which A has a positive status and only exists in relation to its other: 'the other term is purely negatively defined, and has no contours of its own; its limiting boundaries are those which define the positive term' (Grosz 1989: xvi). In other words, 'A' becomes the positive identity in which 'not-A' is measured. The classic examples, for my research, are that the terms 'man' and 'heterosexual' have positive identities, while 'woman' (or not-man) and 'homosexual' (or not-heterosexual) have negative identities. In this logic there is no room for any middle ground or shared identities and this has significance for the construction of hegemonic man/woman identities and spaces because femaleness comes to be constructed as a lack of maleness (Jay 1981; Lloyd 1993; Rose 1993).

Feminist theorist Teresa de Lauretis (1987: 5) argues that: 'the cultural conceptions of male and female as two complimentary, yet mutually exclusive, categories into which all human beings are placed constitute within each culture a gendered system'. The man/woman binary has become known as a natural 'truth' because of the connection between biology, the body and nature. 'Although a child does have a sex from "nature", it is not until it becomes [signified] a boy or a girl that it acquires a gender' (de Lauretis, 1987: 5). Nature within this context appears to exist independently from the cultural inscriptions and representations that later constitute gender.

In summary then, the man/woman binary has been widely considered to be universal and naturally given. Humans are considered to be either male or female and this construction is usually without consideration of space, place, times and cultures. This is a founding distinction of Western culture and its hegemonic status has prompted feminist geographers to outline the implications of the binary. I now offer some examples of research that expose and challenge the man/woman binary.

Challenging the man/woman binary

Feminist geographers have long been concerned with transforming geography's concepts and theories to incorporate gender and sexuality as a fundamental facet of social and cultural spaces. I now ask: what are the costs for geography and for geographers who privilege the masculine over the feminine? One such cost is that men have counted as the 'true' bearers of geographical knowledge. In other words, the gendered character of knowledge has come from the persistent exclusion of women as staff in geography departments. In 1984 the Women in Geography Study Group (WGSG) of the Institute of British Geographers asked 'Why should we, as geographers, be worried about sexism or the "invisibility" of women? Surely things are different these days and women have equal rights with men?' (WGSG 1984: 21). Their book *Geography and Gender: An Introduction to Feminist Geography* charts gendered inequalities and offers critiques of hegemonic masculine geographies of urban spatial structure, industrial location, access to facilities and development. As such, the book offers a challenge to the assumption that cities, location decisions, travel patterns and so on were only made by men. Robyn Longhurst (2001b) notes that the WGSG call for a disruption of geography's theoretical underpinnings. Making women visible was not enough; rather, they insisted that: 'We do not wish this to be seen as merely "adding" women to existing types of geographical analyses without any alteration to the theoretical assumptions underlying these analyses' (WGSG 1984: 254). Susan Smith (2001) remarks that the book is organized around the man/woman binary in that it is concerned with making women visible in human geography and less concerned with disrupting 'rational' thought. For example, missing is a critique of heterosexuality, nor is there any mention of lesbian or queer geographies. The authors, at the time, wished to be taken seriously and perhaps strategically avoided the stereotype that links women–body–sexuality. In retrospect the knowledge that was created by *Geography and Gender* may have been an attempt to align women with rationality. Issues of sexuality and embodiment, at the time, seemed only to reify women as 'nature', thus reasserting the gendered character of the nature/culture binary.

Liz Bondi (1992), in her article '*Gender and dichotomy*', comments on the costs of gendered dichotomous thought. Bondi, like many others before and after her, highlights the longstanding failure of geography departments to appoint women to permanent positions, or to promote those few who were appointed. Bondi (1992) points to surveys conducted in the USA and the UK to reveal that the 'strange case of the missing female geographer' (Zelinski 1973) persisted in the discipline at the beginning of the 1990s.

Linda McDowell (1990) draws attention to the subtle forms of harassment and discrimination that continue to permeate academic institutions. Susan Smith (2001: 256) notes that at the beginning of the new millennium only 6% of professors of geography in Britain are women, and that 'women in geography still struggle with jobs, promotions, ignorance and harassment'.

Gaining legitimacy in the academy has, therefore, been based on the privileging of rationality, objectivity, masculinity. This masculine position can be seen to work in a myriad of ways. Masculinist discourses are evident in, for example, the universalism of knowledge claims, the choice of topics made by geographers. Masculinist rationality is 'a form of knowledge that assumes a knower who believes he can separate himself from his body, emotions, values, past and so on, so that he and his thought are autonomous, context-free and objective' (Rose 1993: 7).

For more than three decades feminist geographers have worked with the binary man/woman – in order to understand particular gendered relations – as well as trying to decouple the binary. Notably, patriarchal spaces have been a prime focus. A useful example is work conducted on geographies of fear. In 1989 Gill Valentine concluded that public places are occupied and controlled by different groups of men at different times. Girls and women felt reasonably safe in public spaces such as parks, city centres and on public transport during the day, but 'as evening draws in it is younger people, particularly men, who are most visible' (Valentine 1989: 388). Other feminist geographers have come to similar conclusions (see Longhurst 1985, 1999; Mehta and Bondi 1998; Pain 1991).

Powerful conceptions of the man/woman binary establish gendered divisions of labour. The intersection of patriarchy and capitalism produces spaces that have been explored by socialist–feminist geographers such as Suzanne Mackenzie and Damaris Rose (1983). They argue that patriarchy genders social relations so that men are expected to undertake waged, productive labour and women are expected to undertake unwaged, reproductive labour. Within this framework, certain spaces in waged labour become gendered as masculine (see Spain 1992) and the home becomes understood as a feminine place. There are many geographical analyses that explore how the man/woman binary is constituted through dominant ideas of home and work, and about public and private. Madigan et al. (1990) suggest that the home is gendered in complex ways that cannot be understood as a simple mapping of male/female onto work/home and public/private concepts. They focus on the ways in which gender influences decisions and ideas about production, exchange and consumption of housing.

Massey (1998) has undertaken research which details and deconstructs dualisms as they relate to certain masculinities in high technology firms. She found that high technology firms were spaces of the mind, of knowledge and reason (scientific rationality and in pursuit of truth) that almost all male employees of these places worked long hours and neglected their family and social lives.

Linda McDowell (1997) has noted that the City of London – and merchant banks – have long been a space dominated by men. She shows that the spaces of merchant banks reflect and reinforce the idealization of the city worker as masculine. Furthermore, feminine bodies are considered to be out of place and 'Othered'. In order for the male merchant bankers to be 'A' or the 'master' subject, they must have the female merchant bankers as 'not-A', and 'Othered'.

Louise Johnson (1989), drawing on the work of Michel Foucault and Elizabeth Grosz, exposes the masculine bias in geography's 'neutral' human subject and offers examples from textile industries in order to show the ways in which bodies become raced, sexualized and Othered through this type of waged labour.

My purpose of outlining this research is not to argue for the valorization of the subordinate pole of the binary – that is, to reverse the hierarchy of man/woman – as this would leave the binary in place. This leaves the masculinist structure of geographical thinking intact and unchallenged (see Rose 1993). The next section provides material that subverts and resists the man/woman binary.

Deconstructing man/woman bodies and spaces

There has been considerable resistance to the binary construction of man/woman and this has been central to feminist geographers' work on embodiment. The body has become a site of intense enquiry, with much of the discussion centred on multiple understandings of masculinity and femininity and on thinking of bodies and spaces in non-dichotomous ways. One such way is to think in terms of what Gillian Rose (1993) refers to as paradoxical space. Thinking of space and bodies in this way means working with the idea of simultaneously occupying centre and margin (masculine and feminine). Bodies can be understood as not just man or just woman, but rather as sexed bodies that are contingent upon time and place.

The notion that all bodies fit into either a male or female has been hotly contested by many feminist geographers. Julia Cream (1995: 31) asks 'What is the sexed body?' and insists that the sexed body is contingent upon time and place. Furthermore, the sexed body could be read as an historical outcome of a range of discourse and meanings centring on biological sex, social gender, gender identity and sexuality. Cream exposes the commonly held belief that we all have to fit into the model of one sex or another. There are many bodies, however, that do not fit the two-sex model. Cream uses examples, such as bodies that have both male and female genitalia, usually known as 'intersexed', transsexuals and people with XXY chromosomes.

Julia Cream is one of many geographers who have become fascinated with embodiment (for a review of this material, see Longhurst 1997) and 'the body' has become a useful resource to articulate, for example, politics of difference. It is now well established that bodies and places are mutually constitutive (Nast and Pile 1998). Within much of this work, 'the call to understand the body is often simultaneously a call for the fluidity of subjectivity, [and] for the instability of binary of sexual difference' (Callard 1998: 388). Postmodernist metaphors such as mobility and fluidity are often used to describe bodies, places and their constitutive relationship. It is still not acceptable, however, to speak of the actual *body fluids* that threaten to break the body's boundaries and mess up other bodies and places (Longhurst 2001a). Bodily fluids are not deemed a legitimate topic in geography because they threaten to mess up clean, clinical, statistical and hard (disembodied and masculinist) geography. When geographers do speak of the body 'they still often fail to talk about a body that breaks its boundaries – urinates, bleeds, vomits, farts, engulfs tampons, objects of sexual desire, ejaculates and gives birth' (Longhurst 2001a: 23). The messy body is often Othered and feminized as a result. What constitutes appropriate issues and legitimate topics to teach and research in geography comes to be defined in terms of reason, rationality and abstraction as though these can be separated out from passion, irrationality, messiness and embodied sensation (Longhurst 2001a).

Some geographers theorizing sexualized bodies, particularly those employing postmodern and critical social theories, tend to posit the body in the realm of ideas, rather than discussing the matter of bodies. Talking about the body is 'safer' than actually embodying our geographies when considering the reality of an often homophobic discipline. Binnie (1997) notes that it is easier to theorize bodies, rather than discuss the gendered and sexualized materialities of bodies and risk getting an academic job or promotions. He argues that there is still squeamishness around sex and sexuality in the production of geographical knowledge. While it may be acceptable to speak of sexuality, for example, it is the actual mention of embodied sexuality (the materiality of sex and sexuality) that creates disciplinary anxieties (Bell 1995). This has certainly been the case when I have presented research on gender and sexuality. My research is occasionally subject to ridicule from colleagues and the popular press. For example, when in 1993 I undertook an Honours research project on the performance and surveillance of lesbian identities and domestic spaces, one staff member did not think my work was 'geography' and wanted the research stopped. It is possible that the tolerance afforded to queer geographers, as noted by Binnie (1997), evaporates when confronted with the materiality of lesbian, gay, bisexual and transgendered bodies. In 1994 and 1995 I researched female body builders and gyms for my Masters thesis. An editor for a national New Zealand newspaper printed a New Zealand Geographical Society advertisement for one of my seminars and placed it under his title: 'It's getting worse folks' (*New Zealand Herald* 1995: 7). Attempts to belittle and marginalize University of Waikato Feminist Geography work on gendered and sexualized embodiment has been documented elsewhere (see Peace et al. 1997). In 2000 at a UK university I presented research on mountains and masculinity (Morin et al. 2001) which was then joked about by some of my colleagues in the departmental tea room. I offer these personal examples to illustrate that the connection between sexed/gendered bodies and geography is often upheld as absurd and humorous.

Despite these risks associated with research on gendered and sexualized bodies many geographers are focusing their studies on embodied performance, bodily displays and the ways in which bodies act out multiple and fluid sexual identities (see the theme issue 'Spaces of Performance, part one and two' in *Environment and Planning D: Society and Space*, 2000). There is some discussion surrounding the construction of gender and race at street festivals and carnival events (see Jackson 1988, 1992; Lewis and Pile 1996; Spooner 1996). Notably, Clare Lewis and Steve Pile (1996) offer complex readings of women's bodies at the Rio Carnival which renders bodies and identity indeterminable.

It is this intersection of sexed/gendered bodies and sexuality which shapes the discussion of the next part of the chapter and I use this work to establish the connections and instabilities associated with man/woman and heterosexuality/homosexuality of marching group bodies at gay pride parades. Before discussing the performativities of sexed and sexualized bodies involved in gay pride parades, I draw on the notion of 'camp' in order to highlight the possibilities of troubling man/woman and heterosexual/homosexual binaries.

Camp places and bodies

Binnie (1997) notes that the notion of a single observable truth has been used to silence and marginalize lesbian and gay bodies and places. He goes on to state that: 'camp can productively work to undermine accepted values and truth, especially the heterosexual definition of space' (Binne 1997: 229). Possibilities of camp are easier to document than to provide a definition as it tends to defy discrete, historical and aesthetic categories. 'Representational excess, heterogeneity and gratuitousness of reference ... signal and contribute to an overall resistance to definition' (Cleto 1999: 3). One of the possibilities of camp is that it can facilitate the creation of non-heteronormative bodies and spaces in oppressive locations. Richard Dyer (1993: 135) states that: 'all the images and words of the society express and confirm the rightness of heterosexuality. Camp is one thing that expresses and confirms being a gay man.' Camp can, I believe, be a useful theoretical possibility to dismantle the static constructions of masculine/feminine and heterosexual/homosexual. This is evident in Judith Butler's discussion of the campness and subversiveness of gay drag:

> The replication of heterosexual constructs in non-heterosexual frames brings into relief the utterly constructed status of the so-called heterosexual original. Thus, gay is to straight *not* as copy is to original, but, rather, as copy is to copy. The parodic repetition of 'the original'..., reveals the original to be nothing other than a parody of the *idea* of the natural and the original. (1990: 31; original emphasis)

The employment of a camp sensibility – a type of appearance and behaviour that mocks and ironizes gender norms – undermines gender assumptions and can create an upheaval of dualisms such as man/woman and heterosexuality/homosexuality in particular cultural contexts.

Geographers have taken up this notion of camp sensibility and the arguments put forward by Butler (1990) to think about the performance of sexual identities in space. David Bell, Jon Binnie, Julia Cream and Gill Valentine (1994: 31) explore the ways in which queer identities are oppositional, transgressing and parodying heteronormativity through a 'long hard look at two current dissenting sexual identities – the hypermasculine skinhead and the hyperfeminine lipstick lesbian'. Bell et al. (1994: 33) make important contributions that trouble gender and upset 'straight' spaces, claiming that the 'mimicry of heterosexuality by gay men and lesbians has the potential to transform radically the stability of masculinity and femininity, undermining its claim to originality and naturalness'. The authors are critical of Butler's work and cite Susan Bordo (1992: 171), who points out that Butler 'does not consider the possibly different responses of different "readers" (male/female, black/white, young/old, gay/straight, etc.) or the various anxieties that might complicate their readings'. Furthermore, Bordo (1992: 171) states that: 'when we attempt to give [Butler's] abstract text some more "body" we run into difficulties'.

In ways that provide a link between the theoretical and empirical matter of this chapter, I include some different audience responses and I focus on the difference that space makes. While Bell et al. (1994) offer detailed case studies of gay skinheads and lipstick lesbians they do not theorize the spaces that constitute these identities. Their article prompted four responses (Kirby 1995; Knopp 1995; Probyn 1995; Walker 1995), in which Probyn (1995: 81) urges geographers to remember 'that the conditions of the production of space as gendered or as sexed are historically, materially and strategically different'. Probyn offers a description of two women in the masculine space of a pub to illustrate the difference that gender and sexuality make and notes that the lesbian subject is always a doubled subject, caught up in being a woman and a lesbian.

I would like to illustrate some of these contradictions and ambivalences surrounding the 'doubled subjects' and places of two gay pride parades: Auckland's HERO Parade, Aotearoa/New Zealand, and the Gay and Lesbian Mardi Gras Parade in Sydney, Australia. Built environments – such as city streets – affect and reflect dominant assumptions about social relations. Heterosexual and familiar relations are assumed 'normal' and those whose bodies and desires do not conform often feel uneasy in spaces structured through heterosexual norms. I have been troubled over the ways in which the gendered/sexed bodies are made and remade at HERO and the Sydney Mardi Gras parades. I have no definitive answers. It seems, however, that their bodies are gendered/sexed in a number of competing and contradicting ways.

Camp performance of masculinity and femininity, at gay pride parades, has the potential to undermine that which is taken for granted and assumed as natural. Butler (1990, 1993) argues that there is no fixed, real or original gender but that the meaning and interpretation of gender is open to change. 'The excessive performance of masculinity and femininity within homosexual frames exposes not only the fabricated nature of heterosexuality but also its claim to originality and naturalness' (Bell et al. 1994: 33). I now turn to the excessive street performances of gay pride parades.

Street performances

The parades are held annually around the months of February and March. These parades are distinct from most northern hemisphere parades, which are held annually around the months of June and July. Northern hemisphere parades tend to be less structured and encourage all people to march in the streets. Members of gay, lesbian, bisexual and transgender communities march to commemorate the New York Stonewall riots of June 1969. These riots have been emblematic of the beginning of the gay liberation movement. Pride, visibility and protest are dominant discourses that construct northern hemisphere parades. Entertainment, parties and performance are defining features of the Sydney Mardi Gras and HERO parades, which helps to attract large audiences.

The particular bodies I wish to focus on are those that perform in marching groups. Marching groups are well known not only in Auckland and Sydney but also in San Francisco, USA. Judging by the response of the audiences watching the parades from the street-side, marching groups, with their synchronized choreographed march/dance routines are the most popular entry in parades. Their routines are a parody of the 'real' female marching teams that are popular in both Aotearoa/New Zealand and Australia (see http://www.marchingnz.org.nz). Marching groups at gay pride parades wear theme-related 'uniforms', for example, the Locker Room Boys at 1995 Sydney Mardi Gras wore gym towels because they wanted to affirm and celebrate 'gym' culture. In 1996 the HERO Marching Boys wore a lifesaving outfit which incorporated swimming shorts, goggles and caps. In 1997 the HERO Marching Boys wore sparkly shorts and narrow sunglasses as a way of adopting a futuristic 'outer space' theme. Marching boys at the 1997 Sydney Mardi Gras Parade wore pink shorts with a heart shape cut-out in the buttocks. Many of the uniforms worn by the marching boys are scanty and expose lots of skin and muscles.

Marching groups tend to have between 40 and 70 members. Their routines involve advancing along the parade route but they also usually have moments when they are stationary and turn directly to face the crowd. Routines tend to be performed to music that is significant for gay male communities, such as 'YMCA' from the group The Village People.

HERO and the Sydney Mardi Gras parades exaggerate the processes by which bodies and places become gendered/sexed and sexualized. At gay pride parades, audiences expect to *see* bodies that defy normative assumptions of gendered/sexed and sexualized bodies, while, and at the same time, they attempt to construct bodies as either masculine or feminine. Gay pride marching teams may be considered as a camp response to the polarization of appearance and reality, of stereotypes and lived experiences and thereby derives its humour from these opposites. Sontag (1966: 276) writes that camp is 'a sensibility that, among other things, converts the serious into the frivolous' and 'one that is alive to a double sense, in which things can be taken ... between the thing as meaning something, anything, and the thing as pure artifice' (Sontag 1966: 281). While Sontag emphasizes the frivolous side of camp, it also has a serious side. Camp may be a means of defiance: a kind of refusal to be overwhelmed by unfavourable odds. It is also a style whereby bodies perform multiple identities, play various parts and assume a variety of roles, both for fun as well as out of political need.

The need to be camp may come from cities that are inscribed with intense heterosexual rigidity. Munt (1995: 115) notes: 'As I became a victim to, rather than a perpetrator of, the gaze, my fantasies of lesbian mobility/eroticism return to haunt me.' Munt argues that in the urban landscape:

> even the protected zones are folding, and yet there are pockets of resistance which pierce the city's metaphoric paralysis with parody: Gay Pride is one such representation, fifty thousand homosexuals parading through the city streets, of every type, presenting the Other of heterosexuality, from Gay Bankers to the Gay Men's Chorus singing 'It's Raining Men', a carnival image of space being permeated by its antithesis. (1995: 123)

Munt (1995) shows that the positions occupied by subjects are never completely fixed. This tension between violence and freedom from oppression is also the topic of Probyn's (1995) response to Bell et al.'s (1994) account of 'lipstick lesbians' in heterosexual space. Probyn states that:

> While one could argue that the sight of two women kissing cannot escape the strictures of heterosexual porn codes, we might also think about, include in our theorizing, the fact that making out in a straight place can be a turn-on. (1995: 81)

Gay pride parades, therefore, become sites/sights of freedom *and* oppression. Parading bodies cannot undo the historicity of the ways in which heterosexism produces a place for the production of the Other. The fact that queer bodies materialize their same sex desire, however, can go some way towards rearticulating that site/sight. It could be argued that the performance of queer desire for heterosexuals is a turn-on. Lustful same sex displays in gay pride parades can be thought of as 'a type of articulation of desire that bends and queers' (Probyn 1995: 81) a (heterosexual) gaze, allowing for a momentarily sexed queer space.

The transgressive possibilities of gender performance may lie somewhere in between the 'author's' intentions and the viewer's (and viewers') intentions. When I spoke to the HERO Marching Boys, it became evident that their routine 'worked' because they played with, and seemed excited about, heterosexuals watching and constructing them as Other. The HERO Marching Boys looked back at the spectators. They also directed sexually suggestive movements and expressions towards the spectators. Several HERO Marching Boys (pseudonyms have been used for research participants) lead the routine and 'turned up the crowd':

> MALCOLM: We're supposed to be the young, young dancing boys.
> BRAD: Sort of turns everybody on. We're supposed to turn up the crowd. (Group interview, 13 February 1996)

The performance of the HERO Marching Boys, in fact, relied on heterosexual audiences. Malcolm and Brad turned up the crowd, and each other, by dancing together and by breaking through the road barriers and dancing with male and female spectators. In these body touching moments, spectators and marching boys disrupt, and hold in tension, male/female and heterosexual/homosexual binaries. The spectators are no longer 'safe' behind the road barriers from queer bodies. Rather, they become part of the spectacle when they are drawn into the performance with the HERO Marching Boys. The actions of the HERO Marching Boys, for a queer sexed space moment, unhinge heterosexism.

Resistance to dominant discourses can also be found by examining the historical construction of heterosexuality. Rather than conceptualize homosexuality and heterosexuality as distinct and separate, another approach is to position homosexuality and heterosexuality as thoroughly entwined (Foucault 1981). This approach insists that heterosexuality depends on homosexuality. Historically, heterosexuality is understood as being derived from homosexuality. According to Foucault (1981), it was not until 1870 that homosexuality was specified as a particular identity category and the concept of heterosexuality could consequently emerge. Heterosexuality, therefore, requires a material conception of homosexuality and the active Othering of homosexuality in order to define itself as a distinct social form (Butler 1990).

Given the structures and complexities of audiences, it is difficult, and not necessarily useful, to attach a single meaning to the marching bodies. It is, however, possible to 'read' parading bodies in several contradictory ways. The construction of the marching bodies relies on the mobility of binaries such as masculine/feminine as they can be understood as performing multiple types of masculinities and femininities.

Performing masculinities

Becoming a Sydney Mardi Gras or HERO marching boy usually involves engaging in specific exercises, usually carried out at a gymnasium, in order to become muscular. Having a muscular body – in Western traditions – has tended to be a masculine pursuit. Body building for men can be seen as the fulfilment of a hegemonic notion of masculinity and/or virility. Male body building can be read as an attempt to render the whole body into the phallus, 'creating the male body as hard, impenetrable, pure muscle' (Grosz 1994: 224). Strength, stamina, control and virility are all attributes associated with muscular male bodies and appear to be sought by the Locker Room Boys' bodies. The connection between muscularity and maleness often becomes naturalized and essentialized.

It is possible to read the bodies of the Locker Room Boys and the HERO Marching Boys as hyper-masculine. A photograph of the HERO Marching Boys appeared in the *New Zealand Herald* (1997: 20) under the caption 'Marcho Men'. The caption 'Marcho' is a play on both macho and marching. The emphasis on the masculine (macho) pertains to the corporeal specificity of their bodies. The photograph of the HERO Marching Boys depicts bodies that are hard, muscular and masculine. The bodies are broad-shouldered and small-waisted. This creates a V-shaped torso that is a traditional marker of masculinity. The lack of clothing (usually the HERO Marching Boys wear small, tight shorts or some variation on this theme) exaggerates the crotch and accentuates stomach, thigh and buttock muscles. These bodies are produced through the 'calculated tearing and rebuilding of selected muscle groups' (Grosz 1994: 143) and through a marching routine that 'pumps' their muscles. The pump is the result of high intensity training and muscle stimulation.

Muscles become engorged with blood and short of oxygen which causes the skin to stretch tight over the muscles. The pump normally results in muscle growth because more blood flushes through the muscles.

The hard, pumped, muscular bodies of the HERO Marching Boys attract a great deal of attention from the crowd. The marching boys move in unison. This gives the appearance of being one large, masculine and muscular body. Attributes of strength, stamina and control become inscribed on the bodies of the HERO Marching Boys during the length of the parade.

Grosz (1994: 202–10) argues that women's bodies tend to be constructed as fluid, uncontrollable and formless. Their bodies are represented as 'leaky' or soft. Female bodies are also represented as threatening, in that they may ooze liquid and disrupt order. Hyper-masculine bodies, however, are thought to remain firmly contained within the epidermis. They are conceived of as the binary opposite of women's bodies. This is not to imply that these male bodies are fluidless. Ironically, their sweating bodies become more masculine as they 'work' their marching routine.

The camp performance of the HERO Marching Boys troubles the polarization of desire, as reported by Bennie:

> The [HERO] Marching Boys were one of the hottest acts of this year's HERO Parade down Auckland's Queen Street. Fans, many of them young heterosexual women, showed their appreciation with enraptured screams. (1995: 1)

This type of hyper-masculinity may help establish a popular profile and glamorous status while at the same time the marching boys can also be conceived of as feminine (but not lacking in status).

Performing feminities

The marching boys devote substantial amounts of time and energy to making their bodies 'right' for gay pride parades. This attention to the appearance and performance of their bodies could be understood as a feminizing process.

The HERO Marching Boys' routine could be interpreted as 'hybrid' activity, part cheerleading and part 'real' marching girl style. Their display, movements and dance can be read as highly 'feminine' and reinforce a camp sensibility. Dyer (1993: 135) claims camp as a survival technique – 'It's being so camp that keeps us going' – while Bell et al. (1994: 36) argue that this is manifested in 'greater care and attention invested in dress, oppositional or otherwise'.

In the *Sydney Morning Herald*'s (1998) feature article on the Sydney Mardi Gras several body tips were disclosed for getting ready to take part in the parade. Under the heading 'Prepping Up' was the following advice to recognize the importance of the event:

> Somewhere between obsession and slovenliness lies a healthy desire to look one's best for an important occasion. And for gay men and lesbians in this town, occasions don't come any more important than Mardi Gras night. (*Sydney Morning Herald* 1998)

Working out at the fitness centre, swimming, running and attention to diet were all recommended to create a body to be proud of. The article also covered procedures for tanning, waxing, tattooing and piercing. Furthermore, shopping for the parade outfit to complement the worked body was also listed as a high priority in the article.

Shops specializing in 'skimpy shimmy shorts' did brisk business in Sydney prior to the Mardi Gras. The bodies being prepped for the parades are reconstituted through traditional feminine practices. Several beauty companies advertised their services prior to the HERO Parade. One company, Adeva Esthetique, specifically targeted gay males in time for the HERO celebrations. Their advertisement read:

> Get off my back this HERO! If you've got hair in places you could do without, why not lose it completely. We offer a safe, hygienic and private waxing/clipping service that'll get the hair off your back and maybe someone new on your tail. Back waxing, Aromatherapy, Facials, Massage. For bookings just give Bradley a call on ... (*express* 1996: 13)

To fulfil the requirements of becoming a member of the hard-bodied marching boys, the gay males also had to become feminized through the effects of training, tanning and waxing. The body becomes marked with the use of makeup, hair products, clothing, posture, diet, weight training and aerobics. Several of the HERO Marching Boys were aerobic instructors.

The HERO Marching Boys team trained for several months prior to the parade. When I observed a training session in Auckland, they had their own female choreographer who regularly trains female ('real') marching girls' teams. Through habitual (feminine) patterns of movement and exercise, the marching boys sculptured their bodies for the parade. The lead marching boy briefs them first. There are lots of jokes about waxing bodies and getting hair done and so on.

Their training and marching routine is designed to accentuate and build their bodies. The performance is high energy and necessarily predicated on exposed flesh. Another reading of the marching boys' bodies is that their small waists, curvaceous pectoral muscles, hairless and oiled flesh are all corporeal indicators of femininity. Furthermore, the process of obtaining a carefully crafted body is a feminizing and disciplining activity. The Sydney Mardi Gras Locker Room Boys are willing to experience pain in order to obtain the ideal body. Here, during the 1995 Mardi Gras Parade, television presenter Elle McFeast interviews one of the Locker Room Boys:

ELLE MCFEAST:	Tell me – how important are gyms to the gay culture?
LOCKER ROOM BOY:	We are the gay culture, we are the essence of the gay culture.
ELLE MCFEAST:	A lot of waxing has gone into preparations?
LOCKER ROOM BOY:	Yes, we have gone to a lot of trouble, a lot of effort, a lot of pain, to make people happy. (Australian Broadcasting Corporation 1995)

The evocation of femininity (through training, routines, prepping and so on) enables the bodies to be built to a 'parading' state of hyper-masculinity. As a result, these marching boys' bodies also confuse traditional or 'natural' corporeal indicators of masculinity.

There are moments of contestatory politics and other paraders have reacted to the status of the official marching boys. One such reaction was a parade entry called the *express* Check-Out Chicks. Robert, a member of this group, described their intentions:

> We are twelve drag queens or so, and it's almost a parody on the [HERO] Marching Boys and it's a full parody of like, like, normality. We're taking like Foodtown [supermarket] check-out trolleys and marching. (in-depth interview, 7 February 1996)

The performance of the HERO Marching Boys was now subject to another performance. Like the HERO Marching Boys, the Check-Out Chicks used whistles in their routine, but rather than 'strip down', like the HERO Marching Boys, the *express* Check-Out Chicks 'dressed up' in full drag. They explicitly hyper-feminized themselves in an attempt to send up the hyper-masculinism of the HERO Marching Boys. The supermarket trolleys also feminized their performance and enhanced the parody. The name 'Express' was a play on their sponsor, a newspaper called *express: new zealand's newspaper of gay expression*. 'Express' is also a reference to the express, or fast-track check-out lanes at the Foodtown supermarkets. Shopping at supermarkets can also be understood as a feminized activity, or what Robert calls 'normal' activity.

There are two things worth noting here. The first is that spectators are not expecting to see 'normal' or 'everyday' activities such as shopping in a gay pride parade, which may upset their notions of the Other. The second point that is worth noting is that this is a parodic repetition of a parodic repetition – in other words, this performance is a parody of a parody, in that the regulatory forces that shape marching groups, are camped up by the HERO Marching Boys' bodies, and in turn, restyled by the *express* Check-Out Chicks. One outcome of this might be that any sense of an original gender performance is thoroughly misplaced and reimagined. 'Understood at least partially as a political act, camp drag produces an effect that will critically comment on or at least inflect the gender system of a particular historical/cultural moment in unpredictable, but readable, ways' (Piggford 1999: 288). Marching Boys play with their respective cultures as well as with gay pride audiences. In this way, gendered and sexualized bodies are produced in relation to each other, as well as in relationship to the queered spaces of gay pride parades.

Concluding remarks

The aim of this chapter was to examine the gendering of geographical knowledge and the ways in which bodies and places are both subject to and resist gendered and sexualized dualisms. I have outlined the history, spatial and cultural contexts that work to establish the powerful and hierarchical binary discourse of man/woman. The man/woman binary has been crucial to the production of hegemonic knowledge in several ways. Feminist geographers have commented on the exclusion of women in geography departments. The early identification of inequality led to feminists finding a voice by examining the processes and spatial implications of gendered binaries. Feminists worked with these binaries in order to highlight the inequalities that the binaries sustain. The masculinist rationality of the discipline has been subject to critical scrutiny and as such diversity and difference now dominate the theoretical agenda. Gender cannot be considered a stable entity, rather gender is performed in different ways in different spaces, bodies and times.

In my own research I have adopted notions of camp and some of Butler's (1990, 1993) work to highlight the performativity of gendered and sexualized bodies and spaces in order to trouble static and binary categories of gender and sexuality. Over the past decade there has been a persistent call to examine alternative frameworks and discourses for capturing the performativity, fluid, diverse and contradictory ways in which sexed and sexualized bodies are made. My account of parades and bodies may be one way to do this.

I have teased out some of the ways in which marching bodies are gendered/sexed at two particular southern hemisphere parades. Bodies on parade, such as the HERO Marching Boys and *express* Check-Out Chicks in Auckland and the Locker Room Boys of the Sydney Mardi Gras Parade, demand a rethinking of the terms of sexual differences and gendered bodies. By reading the camp performance of marching bodies as masculine, feminine and as *both* masculine and feminine, I hope I have shown bodies to be volatile, mobile and contradictory. The deployment of particular identities serves to undermine the naturalness and originality of gendered bodies and to reveal gender as performative. The binary man/woman becomes a regulatory fiction when queer bodies camp up notions of femininity and masculinity. Social meanings produced in these parades – in which the paraders, who are both shaped by these institutions and are agents of change – produce changes that may both serve hegemonic interests and challenge existing power relations.

Throughout this chapter I have raised epistemological issues in camp geographies in order to cause trouble for the man/woman binary. Geographers such as Binnie (1997) and Bell et al. (1994) provide a platform for the ways in which I have been able to trouble gender and sexuality. I have argued for a consideration of places and bodies as camp in order to trouble the regulatory practices of heteronormativity. I have presented a case study on marching bodies in gay pride parades but I would also like to return to the cat toys Big Yellow Bob and Swanky Ken Dog. Like the marching bodies, these toys challenge popular held assumptions about gender and sexuality because they parody gender heterosexual identities *and* homosexual identities. Their bodies become readable in a myriad of ways that may create anxieties and, hopefully, some kitty hoots too.

FURTHER READING

Bondi, Liz (1992) 'Gender and dichotomy', *Progress in Human Geography*, 14 (3): 438–45
This is an early and insightful commentary on geography and binaries.

Lloyd, Genevieve (1993) *The Man of Reason: 'Male' and 'Female' in Western Philosophy.* London: Routledge
Lloyd provides historical reasons and philosophical background of gendered dualisms.

Binnie, Jon (1997) 'Coming out of geography: Towards a queer epistemology?', *Environment and Planning D: Society and Space* 15: 223–37
Binnie offers a queer challenge for geographers in order for them to disrupt dichotomous and heterosexist epistemologies. In doing so he subverts the ways in which geography has rested on gendered dichotomies.

Bell, David, Binnie, John, Cream, Julia and Valentine, Gill (1994) 'All hyped-up and no place to go', *Gender, Place and Culture: A Journal of Feminist Geography* 1: 31–48
A good example of the performance of gendered and sexualized bodies and spaces.

Massey, D. (1996) 'Masculinity, dualisms and high technology', in Nancy Duncan (ed.), *BodySpace.* London: Routledge. pp. 109–26
A commentary on binaries, masculinity and high technology.

For research that upsets hegemonic and static understandings of embodiment, see:
Cream, Julia (1995) 'Re-solving riddles: The sexed body,' in David Bell and Gill Valentine (eds), *Mapping Desires: Geographies of Sexualities.* London: Routledge. pp. 31–40

Johnston, Lynda (1998) 'Reading sexed bodies in sexed spaces', in Heidi Nast and Steve Pile (eds), *Places Through the Body.* London: Routledge. pp. 244–62

Longhurst, Robyn (2001) *Bodies: Exploring Fluid Boundaries.* London and New York: Routledge

For further discussions of camp and queer theories, see:
Cleto, Fabio (ed.) (1999) *Camp: Queer Aesthetics and the Performing Subject: A Reader.* Edinburgh: Edinburgh University Press

References

Australian Broadcasting Corporation (1995) *Sydney Mardi Gras Parade* (programme transcript). ABC Video, Sydney.

Bell, David (1995) '[Screw]ing geography (censor's version)', editorial in *Environment and Planning D: Society and Space*, 13 (2): 127.

Bell, David, Binnie, Jon, Cream, Julia and Valentine, Gill (1994) 'All hyped-up and no place to go', *Gender, Place and Culture: A Journal of Feminist Geography*, 1: 31–48.

Bennie, J. (1995) 'Warriors court marching boys', *Man to Man: New Zealand's National Gay Community Newsletter*, 13 April: 1.

Berg, Larry (1994) 'Masculinity, place and a binary discourse of "theory" and "empirical investigation" in human geography of Aotearoa/New Zealand', *Gender, Place, Culture: A Journal of Feminist Geography*, 1 (2): 245–260.

Binnie, Jon (1997) 'Coming out of Geography: Towards a queer epistemology?', *Environment and Planning D: Society and Space*, 15: 223–237.

Bondi, Liz (1990) 'Feminism, postmodernism and geography: Space for women?', *Antipode*, 22 (2): 156–167.

Bondi, Liz (1992) 'Gender and Dichotomy', *Progress in Human Geography*, 14 (3): 438–445.

Bondi, Liz and Mona, Domosh (1992) 'Other figures in other places: on feminism, postmodernism and geography', *Environment and Planning D: Society and Space*, 10: 199–213.

Bordo, Susan (1986) 'The Cartesian masculinization of thought', *Signs*, 11: 239–256.

Bordo, Susan (1992) Review Essay: 'Postmodern subjects, postmodern bodies', *Feminist Studies*, 18: 159–175.

Butler, Judith (1990) *Gender Trouble: Feminism and the Subversion of Identity*. New York and London: Routledge.

Butler, Judith (1993) *Bodies that Matter: On the Discursive Limits of Sex*. New York and London: Routledge.

Callard, Felicity (1998) 'The body in theory', *Environment and Planning: Society and Space D*, 16: 387–400.

Cavendish, A.P. (1964) 'Early Greek philosophy', in D.J. Connor (ed.), *A Critical History of Western Philosophy*. New York: The Free Press. pp. 1–13.

Cleto, Fabio (ed.) (1999) *Camp: Queer Aesthetics and the Performing Subject: A Reader*. Edinburgh: Edinburgh University Press.

Cream, Julia (1995) 'Re-solving riddles: The sexed body', in David Bell and Gill Valentine (eds), *Mapping Desires: Geographies of Sexualities*. London: Routledge. pp. 31–40.

de Lauretis, Teresa (1987) *Technologies of Gender: Essays on Theory, Film, and Fiction*. Bloomington: Indiana University Press.

Dyer, Richard (1993) 'Straight acting', in Richard Dyer (ed.), *The Matter of Images: Essays on Representations*. London: Routledge. pp. 133–136.

express: new zealand's newspaper of gay expression, 1 February 1996: 13.

express: new zealand's newspaper of gay expression, 21 February 1998.

Foucault, Michel (1981) *The History of Sexuality. Volume One: An Introduction* (trans. R. Hurley). London: Penguin Books.

Grosz, Elizabeth (1989) *Sexual Subversions: Three French Feminists*. Sydney: Allen and Unwin.

Grosz, Elizabeth (1994) *Volatile Bodies: Towards a Corporeal Feminism*. St Leonards: Allen and Unwin.

Jackson, P. (1988) 'Street life: The politics of carnival', *Environment and Planning D: Society and Space*, 6 (2): 213–227.

Jackson, Peter (1992) 'The politics of the streets: A geography of Caribana', *Political Geography*, 11: 1–22.

Jay, Nancy (1981) Gender and dichotomy. *Feminist Studies*, 7: 38–56.

Johnson, Louise C. (1989) 'Embodying geography: Some implications for considering the sexed body in space', *New Zealand Geographical Society Proceedings of the 15th New Zealand Geography Conference*, Dunedin. pp. 134–138.

Johnson, Louise C. (1994) 'What future for feminist geography?', *Gender, Place and Culture: A Journal of Feminist Geography*, 1 (1): 103–113.

Johnston, Lynda and Valentine, Gill (1995) 'Where ever I lay my girlfriend that's my home: Performance and surveillance of lesbian identity in domestic environments', in David Bell and Gill Valentine (eds), *Mapping Desires: Geographies of Sexualities*. London: Routledge. pp. 99–113.

Johnston, Lynda (1996) 'Pumped up politics: Female body builders refiguring the body', *Gender, Place and Culture: A Journal of Feminist Geography*, 3 (3): 327–340.

Johnston, Lynda (1997) 'Queen(s') Street or Ponsonby Poofters? The embodied HERO Parade site', *New Zealand Geographer*, 53: 29–33.

Johnston, Lynda (1998) 'Reading sexed bodies in sexed spaces', in Heidi Nast and Steve Pile (eds), *Places Through the Body*. London: Routledge. pp. 244–262.

Johnston, Lynda (2001) '(Other) bodies and tourism studies', *Annals of Tourism Research: A Social Science Journal*, 28: 180–201.

Johnston, Lynda (2002) 'Borderline bodies at gay pride parades', in Liz Bondi, Hannah Avis, Amanda Bingley et al. *Subjectivities, Knowledges and Feminist Geographies. The Subjects and Ethics of Social Research*. Lanham, MD: Rowman and Littlefield.

Kirby, Andrew (1995) 'Straight talk on the homo question', *Gender, Place and Culture: A Journal of Feminist Geography*, 2: 89–96.

Knopp, Lawrence (1995) 'If you're going to get all hyped up you'd better go *somewhere*!', *Gender, Place and Culture: A Journal of Feminist Geography*, 2: 85–88.

Lewis, Claire and Pile, Steve (1996) 'Woman, body, space: Rio Carnival and the politics of performance', *Gender, Place, and Culture: A Journal of Feminist Geography*, 3: 23–41.

Longhurst, Robyn (1985) *Sexual violence: One of the reasons why space, both public and private, belongs to men*. Human Geography Occasional Paper 3, University of Waikato.

Longhurst, Robyn (1995) 'Geography and the body', *Gender, Place and Culture: A Journal of Feminist Geography*, 2: 97–105.

Longhurst, Robyn (1997) '(Dis)embodied geographies', *Progress in Human Geography*, 21 (4): 486–501.

Longhurst, Robyn (1999) 'Gendering place', in Richard Le Heron, Laurence Murphy, Pip Forer and Margaret Goldstone (eds), *Explorations in Human Geography: Encountering Place*. Auckland: Oxford University Press. pp. 151–172.

Longhurst, Robyn (2001a) *Bodies: Exploring Fluid Boundaries*. London and New York: Routledge.

Longhurst, Robyn (2001b) 'Classics in human geography revisited', *Progress in Human Geography*, 25 (2): 252–255.

Lloyd, Genevieve (1993) '*The Man of Reason: "Male" and "Female" in Western Philosophy*'. London: Routledge.

Mackenzie, Susan and Damaris, Rose (1983) 'Industrial change, the domestic economy and home life', in J. Anderson, S. Duncan, and R. Hudson (eds), *Redundant Spaces in Cities and Regions?* London: Academy Press. pp. 155–200.

McDowell, Linda (1990) 'Sex and power in academia', *Area*, 22: 323–332.

McDowell, Linda (1991) 'The baby and the bath water: Diversity, deconstruction and feminist theory', *Geoforum*, 22: 123–133.

McDowell, Linda (1997) *Capital Culture: Gender at Work in the City*. Oxford: Blackwell.

Madigan, R., Munro, M. and Smith, S. (1990) 'Gender and the meaning of home', *International Journal of Urban and Regional Research*, 14: 625–647.

Massey, D. (1998) 'Masculinity, dualisms and high technology', in Nancy Duncan (ed.), *BodySpace*. London: Routledge. pp. 109–126.

Mehta, Anna and Bondi, Liz (1999) 'Embodied discourse: on gender and fear of violence', *Gender, Place and Culture: A Journal of Feminist Geography*, 6 (1): 67–84.

Morin, Karen with Longhurst, Robyn and Johnston, Lynda (2001) '(Troubling) spaces of mountains and men: New Zealand's Mount Cook and Hermitage Lodge', *Social and Cultural Geography*, 2 (2): 117–139.

Munt, Sally (1995) 'The lesbian flâneur', in David Bell and Gill Valentine (eds), *Mapping Desire: Geographies of Sexualities*. London: Routledge. pp. 114–125.

Nast, Heidi and Pile, Steve (eds) (1998) *Places Through the Body*. London and New York: Routledge.

New Zealand Herald (1995) 'Harold Angel week's end: it's getting worse folks', 19 August: 7.

New Zealand Herald (1997) 'Marcho Men', 24 February: 20.

Pain, Rachel (1991) 'Space, sexual violence and social control: integrating geographical and feminist analyses of women's fear of crime', *Progress in Human Geography*, 15: 415–431.

Peace, Robin with Longhurst, Robyn and Johnston, Lynda (1997) 'Producing feminist geography "Down Under"', *Gender, Place and Culture: A Journal of Feminist Geography*, 4 (1): 115–119.

Piggford, George (1999) 'Who's that girl? Annie Lennox, Woolf's Orlando, and female camp androgyny', in Fabio Cleto (ed.), *Camp: Queer Aesthetics and the Performing Subject: A Reader*. Edinburgh: Edinburgh University Press. pp. 283–299.

Probyn, Elspeth (1995) 'Lesbians in space: Gender, sex and the structure of missing', *Gender, Place and Culture: A Journal of Feminist Geography*, 2: 77–84.

Rose, Gillian (1993) *Feminism and Geography: The Limits of Geographical Knowledge*. Cambridge: Polity Press.

Smith, Susan (2001) 'Classics in human geography revisited', *Progress in Human Geography*, 25 (2): 255–260.

Sontag, Susan (1966) *Notes on Camp: Against Interpretation*. New York: Delta.

Sontag, Susan (1983) *A Susan Sontag Reader*. London: Penguin.

Spain, Daphne (1992) *Gendered Spaces*. Chapel Hill, NC: University of North Carolina Press.

Spooner, Rachel (1996) 'Contested representations: Black women and the St Paul's carnival', *Gender, Place and Culture: A Journal of Feminist Geography*, 3 (2): 187–203.

Sydney Morning Herald (1998) 'City Search Guide to the Mardi Gras', (http://sydney.citysearch.com.au/E/F/SYDNE/0000/00/64/) 3 March.

Television 3 Network Services Limited (1997) *The 1997 HERO Parade* (programme transcript). Television 3, Wellington.

Valentine, Gill (1989) 'The geography of women's fear', *Area*, 21 (2): 385–390.

Walker, Lisa (1995) 'More than just skin deep: fem(me)ininity and the subversion of identity', *Gender, Place and Culture: A Journal of Feminist Geography*, 2: 71–76.

Women and Geography Study Group of the Institute of British Geographers (1984) *Geography and Gender: An Introduction to Feminist Geography*. London: Hutchinson and Explorations in Feminism Collective.

Zelinski, Wilbur (1973) 'The strange case of the missing female geographer', *Professional Geographer*, 25 (2): 101–105.

8 Nature: Culture

Michael Watts

> It is evidently quite possible to cut off one's own hand without feeling pain. People whose hand has become trapped in machinery have sometimes amputated it painlessly ... On the other hand if you smack a child really hard in jest, he is quite likely to cry even so. Meaning can mould physical responses, but they are constrained by them too. The adrenal glands of the poor are often larger than those of the rich, since the poor suffer more stress, but poverty is not able to create adrenal glands where none exist. Such is the dialectic of nature and culture.
>
> (Eagleton 2000: 87)

> Neither the frontiers between the wild and the cultivated, nor those that lie between the past and present, are so easily fixed ... we need only to poke below the topsoil ... to discover an obstinately rich loam of memory ...
>
> (Schama 1995: 574)

It has been said, with good reason, that nature and culture are perhaps the two most complex words in the English language. They carry, as Raymond Williams (1976) once said, an enormous amount of human history. Taken together, the two words are often assumed to be opposites – the material and the ideal, the biological and the semiotic, a realm of law and a world of contingency – but on closer examination their polarities are tangled, difficult and intractable. It is fashionable today, for example, to think of nature as derived in some way from culture – the wilderness as an idea or the tropical rainforest as a way of writing or cultural representation (see Slater 2002; Stepan 2001) – but in an etymological sense, culture is *derived* from nature. One of its original meanings was husbandry or natural growth. Francis Bacon even wrote of the culture and manuring ('manurance' was the word he deployed) of human minds! But the Latin root (*colere*) also conveys other meanings: habitation, worship, protection. As Terry Eagleton puts it, culture inherits the imposing mantle of religious authority but also has 'uneasy affinities with occupation and invasion' (2000: 2). The same might be said of nature and its historical semantics.

These words obviously refer to different things; they have different referents. A tropical rainforest, the double-helix or a death-drive are clearly distinct, in some profound way, from a Bororo fishing ceremony, a Catholic ritual or a Mayan hieroglyph. They are surely not reducible to one another. But the referents themselves are not stable. They are shifty, tangled and mixed up; antinomies seem to proliferate. Nature may be culture's other but there is much traffic, a veritable information highway, between them. If nature is in some sense cultural, culture is in turn built out of the 'ceaseless traffic with nature which we call labour' (Eagleton 2000: 4). It is for this reason that David Harvey (1996: 187) can comfortably claim that world cities like New York or Mumbai are wholly natural: raised out of wood, sand, iron, stone and the like. They are of nature.

What, then, are the semantic distinctions and instabilities within and between nature and culture? How might we grasp their historical semantics? Raymond Williams (1976) has charted three major modern senses of the term culture: as utopian critique, as a way of life, and as artistic creation. The first must be situated against the horrors of industrial capitalism, as a sort of anti-capitalist impulse. The word first comes to mean civility and later civilization (understood as progress), but around the turn of the nineteenth century the descriptive and normative aspects of the word, as Eagleton says, 'fly apart' (2000: 10). Civilization appears increasingly debased, predatory and culture appears as a response to 'radical and painful change' (Williams 1993: 60). The second shift marks culture as a way of life and is implicitly a critique of the universalism of the Enlightenment. Culture is rendered multiple to encompass, for example, the Romantic anti-colonial desire to reclaim the exotic and the traditional in the face of industrial triumphalism. Cultures are organic and rooted and could furnish, in the hands of Johan Herder or Edmund Burke, a radical critique of Eurocentrism or the notion of industrial capitalism as universal civilization. And third, Williams charts culture as a gradual specialization to the arts whether broadly construed as intellectual pursuits or science, or more narrowly confined to literature and the imaginative arts. Culture figures here, above all else, as refinement, impartiality and 'agreeable manners and an open mind' (Eagleton 2000: 180).

In its modern iterations culture appears, in Williams' account, as an improper word, 'both synonymous with the mainstream of western civilization and antithetical to it' (Young 1995: 53). Culture's polysemic qualities are perfectly captured in a new geography handbook (Anderson et al. 2003): culture, the authors point out, can be a distribution of things, a way of life, a universe of meaning, a way of doing and a field of power. In Williams' language culture is a 'binding' word which sutures these differing activities and their interpretation, and is also what he calls an 'indicative' word appearing in specific forms or lineages of thought (for example nationalism or racism). These complex and shifting modern meanings cannot be identified, so to speak, in advance; rather they are philosophical and historical texts, always deployed in particular ways, and inextricably bound up with the problems they are being used to discuss.

Nature might seem less immune to the semantic complexity of culture. One might start with Kate Soper's (1995) exemplary meditation on the subject, *What is Nature?* Nature, she says with admirable clarity, refers to 'those material structures and processes that are independent of human activity (in the sense that they are not a humanly created product), and whose forces and causal powers are the necessary condition of every human practice' (1995: 132–3). But there are those – post-structural theorists like Stanley Fish or Richard Rorty – who would deny such independence. It is culture, they say, all the way down. To deploy the mantra of the social constructivists, 'Nature cannot pre-exist its construction' (Haraway 1992: 296). But in any case, Soper's definition on its face says little about the manifold ways in which the word is, and has been, deployed. Rather alarmingly for our purposes, Raymond Williams has awarded the accolade of 'most complex word' to none other than nature! Soper is aware of this complexity – she begins her book with a homage to Williams – and indeed her study turns on two differing 'politics of nature', two contrasting views of nature. One ('nature-endorsing') emerges as a response to the destruction of the natural world and seeks respect and conservation for the environment; the second ('nature-sceptical') focuses on the ideological functions to which nature (as a representation or construction) is, or can be, put. Both of these perspectives can be accommodated within Williams' (1976: 219–24) three areas of meaning of the word nature: as the essential quality or character of something (it is natural to truck, barter and exchange), as the inherent force which directs the world or human beings or both (Freudian drives), and the material world itself which may or may not include humans (the laws of the global ecosystem). All three meanings to some extent overlap, and have been in use since the thirteenth to seventeenth centuries. They are all alive and well today. Like culture, nature started its semantic life describing a quality or process and subsequently became an independent noun. A fundamental part of this transformation is the semantic shift from nature as a specific singular to an abstract singular (from nature to Nature one could say). It is, as Williams says (1976: 220), 'structurally and historically cognate with the emergence of God from a god', presupposing in other words a single prime cause. Any account of the differing conceptions of nature in the West would encompass a vast swath of human thought, but historically each articulation of nature would be a 'powerful personification' as Williams puts it – as goddess, as minister, as monarch, as lawyer, as natural selector. Each conception is typically static and driven by a set of laws. The very notion of personification makes the point that all ideas of nature are always the work of flesh and blood persons.

Culture and nature are, in sum, keywords, each of which carries, over long periods, many of the variations of human thought. The meanings are multiple, shifting and unstable. And as Williams makes clear, they are sites or objects of political conflict. The complex of senses attributed to culture, says Williams, 'indicates a complex argument about the relations between general human development and a particular way of life ...' (1976: 81). The same might be said of nature: its complex senses indicate a struggle over the relations between an abstract singular nature and particular ways of life. A part of the struggle and conflict over the meanings of keywords is what connects the different meanings for each word. In the case of culture the meanings are surely linked by the sense of *failure* of culture as civilization (see Eagleton 2000); it provides a critique of the (unquestioned) great unfolding of human progress and self-improvement. For nature the different meanings are linked to the common identification of nature with reason, and the possibility of contrasting the is (what exists) with the ought (what should or could be): the state of nature with the state of society. The Enlightenment and the Romantic movements were two such responses to the perceived dissonance between nature and what man had made of himself. Underlying each was a search for order, law and a universal, inherent and primary force.

Distant riders: tracing nature/culture

It is no accident that words like nature and culture, given their centrality to Western thought, have a ubiquity across the social and human sciences, and this is no less the case for geography. In his magisterial study *Traces on the Rhodian Shore* (1967), Berkeley geographer Clarence Glacken charted, over the period between the ancients and the end of the eighteenth century, the parallel histories in Western thought of three foundational ideas through which nature and culture came to be interpreted: the idea of a *designed earth*, the *influence of the environment on society*, and the role of humans *as modifiers of the natural world*. This classic *tour d'horizon* of Western ideas of nature and culture is by any measure a towering achievement. In the canon of environmental history and cultural geography it is a, perhaps *the*, foundational text. According to the *New York Times*, *Traces* is one of the most important books published in the English language since the end of World War II. Glacken began with the Hellenistic age – the affinity of theology and geography, Xenophon's treatise on design, Aristotle's teleology of nature – and for each epoch explored the ways in which a panoply of great thinkers wrestled with the idea of the earth as a planned abode, of environmental influence and determinism within a divinely created world, and of the capacity of humans to control nature through technology, labour power and science. Schooled in intellectual history (he saw himself as a continuator of the founding generation of the 'history of ideas' movement, in particular of Arthur Lovejoy), Glacken sought to emphasize, and it is an enduring strength of the book, the historical continuities of ideas. Ideas and bodies of thought are 'persistent': 'large related bodies of thought appear, at first like distant riders stirring up modest dust clouds, who, when they arrive, reproach one for his slowness in recognizing their numbers, strength and vitality' (1967: xi).

Glacken's foundational nature/culture themes were typically modified and enriched by other theories relating to culture, history and the nature of the earth. Plenitude, the organic analogy, the role of human institutions such as religion and government, for example, all figured prominently in the ways in which a designed earth or human-induced transformation of nature was articulated in different places and at different times. His account stopped, however, at the very birth of industrial modernity in the late eighteenth century when, as he put it:

> [T]here ends in Western civilization an epoch in the history of man's relationship to nature. What follows is of *an entirely different order*, influenced by the theory of evolution, specialization in the attainment of knowledge, acceleration in the transformation of nature. (1967: 705; emphasis added)

The Industrial Revolution, a term Glacken disliked, marked a watershed after which the ideas of design, influence and modification carried entirely new meanings. Much of the sense of rupture – ideological, scientific, political economic, cultural – wrought by industrial capitalism was foreseen in the ideas of Malthus, Godwin and Buffon, three figures selected by Glacken to close out his encomium. *Traces* was seen by Glacken as a prelude for a subsequent treatment of such towering figures as Marx, Darwin, Weber, Durkheim and Einstein as he brought his story up to the arrival of the nuclear age – and in so doing to return to the very old idea of apocalypse and extermination in Western thought. A sequel to *Traces* was in fact completed in the early 1980s, continuing his extraordinary intellectual history up through the nineteenth and twentieth centuries, but the manuscript was mysteriously destroyed.[1]

Glacken's challenge, to chart the vast complexity and the dense intellectual traffic between nature and culture from antiquity to the birth of modernity, rested on his abiding concern with 'great ideas' and intellectual biography. Yet in spite of his historical range and depth, he was less sensitive both to the ways in which particular knowledges and ideas were produced and to the material conditions of their production – what we would now call discourse and discursive practice – and to the epistemological significance of the nature/culture binary. In many respects, then, Glacken represents a sort of paradox with respect to the historical semantics of culture and nature with which we began. On the one hand, he is acutely aware of the shifting meanings of the words and within these shifts some obdurate and persistent commonalities over time (always shaped by local circumstance). On the other, he is less willing to engage with the deployments of the words in relation to what Williams called political conflict and the relations between meanings. One can see this in, for example, his treatment of Thomas Malthus, who in many respects is a key figure in the history of the relations between nature, culture and modernity. Glacken's treatment of Malthus is thorough but quite anodyne. Malthus, he says, brings to maturity a line of thinking (Montesquieu, Wallace and Hume among them) in which nature imposes limits. Glacken points out that Malthus likened the earth to a closed room, an island and a reservoir (1967: 639). One aspect of nature—land—was especially key for Malthus because its productivity was constantly outrun by the 'passion of the sexes', the tendency for human numbers to increase geometrically, to outstrip the means of subsistence and to produce, famine, war, vice and poverty. His 'principle of population' represented the diety acting through the laws of nature. And as Glacken properly observed of Malthus' view (1967: 645), the laws of nature are benevolent if men know their operations and abide by them. Glacken notes that Malthus 'did not believe that any improvement in human nature could be expected or that the reform in government or institutions could – or should – alter the operations of the principle of population' (1967: 644). It is here that Malthus' debate with Godwin was so important, as Glacken observes. All of this is quite helpful in seeing how a key figure thinks about nature and culture through the principle of population and natural law, but in a curious way Glacken misses what the debate over Malthus and his ideas was about. What was the work – discursive and practical – that nature and culture performed in Malthus' corpus?

Malthus was born in 1766, the son of a liberal English country gentleman whose liberalism – Rousseau and his mistress dined at the family house, the Rookery, near Dorking – he bequeathed to Thomas through an education within the dissenter's tradition. He graduated in mathematics from Jesus College, Cambridge and was subsequently elected a Fellow in 1793, the same year that he took orders in the Anglican Church. He published his second essay on population (vastly expanded and less polemical) in 1803, the same year he became Rector of Walesby in Lincolnshire. Malthus' personal influence was probably at its height in Britain in the second decade of the nineteenth century when the debate over the abolition of the Poor Laws was at its most intense, but the suturing of his name to any discussion of demographic growth and poverty has persisted in a way that exceeds the legacies of many of his great contemporaries – Adam Smith, David Ricardo, David Hume, William Godwin and Tom Paine.[2] Malthus was manifestly staid in his life and behaviour and made every effort to present his ideas as a model of balanced and reasonable argument – what we might now call logical empiricism.

One must be careful of seeing Malthus simply as a voice of political reaction, or some simple mouthpiece of the landed gentry, or as a pessimist, which is quite unhelpful.[3] All of these judgements are of course coloured by the fact that his ideas – to be generous – 'evolved' and changed over time.[4] All of which is to say that it is both true, and yet incomplete and overly simplified, to suggest that Malthus' views on population and poverty were derived from the vantage point of landed property or that they simply endorse bourgeois interest. Malthus was a voice of *reaction* – recognizing that his reactionary views necessitated reform, and indeed a quite radical reform by which the moral economy of the eighteenth century was displaced by new forms of liberal governance and forms of personal responsibility. Malthus' concern with political liberty is, as with all of his work, seized by a sort of paternalism and an irreducible class fear.

Malthus can be approached from a quite different perspective than that of population narrowly understood if we begin with Karl Polanyi's germinal book *The Great Transformation* (1967) and his account of the setting in which Malthus emerged as a public intellectual. Polanyi noted that the debates over Speenhamland (the Poor Law reforms following the hardships of the early 1790s) – what he called pauperism and political economy – 'together form a great whole – the discovery of society' (1967: 103). The figure of the pauper was in other words part of a conversation – in which Malthus was a towering figure – in which poverty was constituted along with a form of liberal governance (Dean 1991). In a sense then population *per se* was less central than the shift from an Elizabethan moral economy to a liberal mode government, fostering a space of individual autonomy and minimal state action. The conversation represented a break with 'the discourse of the poor' as Mitchell Dean calls it (1991: 18) along three dimensions:

1 The emergence of a strong abolitionist strain toward the Poor Laws.
2 The insistence on voluntary charity as morally superior to the compulsion inherent in the poor rates.
3 The substitution of make-work schemes by various contributory methods (life annuities, insurance plans and mutual assistance).

Implicit in the shift from Speenhamland in 1795 to the repeal of the Poor Laws in 1834 is a parallel tectonic movement, namely a redefinition of the field of action of the state: a withdrawal of the state responsibility for relief to categories of the propertyless, and the centralization and bureaucratization of the state apparatus of relief. To the extent that this was part of a sometimes acerbic debate over ethics, the historic role of Malthus was to root them in the overwhelming bioeconomic laws of population and subsistence.

The figure of the pauper and political economy was of course, as Lord Cockburn put in 1868, 'soaked in one event' – the revolution in France. The Malthusian question and the poverty debate can only be grasped then as one contentious pole within the political arena of what I shall call, following David McNally (1995), 'popular radicalism'. Paine's *The Rights of Man* appeared in 1791 and it is to be recalled that it was enormously popular, selling 200,000 copies within a year. *The Rights of Man* politicized poverty – it accepted the logic of the free market but provided a programme for political reform. The emergence of two radical societies in Britain – the Constitutional Society and the London Corresponding Society – can be seen as part of a stream of radical English Jacobinism, a continuous revolutionary-republican underground which runs from the mid 1790s to Chartism. The dearth of 1795–6 and the riots and grain seizures which it generated marked a forward surge of this Jacobin sentiment. At the heart of the popular radicalism and its opposition to political economy was a debate within the English ruling class between those who sought to buttress traditional obligations to the poor as a means of circumventing mass upheaval, and those who favoured jettisoning such self-defeating efforts in place of the abolition of the Poor Laws (and in effect new institutionalized forms of personal responsibility and an effective national labour market). Malthus enters into this fray quite explicitly. It is to be recalled that his Essay was addressed explicitly to Condorcet and Godwin (and later Robert Owen and the Spenceans), two utopian/anarchist/socialists for whom the perfectibility of humankind lay in the possibility of the social reform of institutions. Malthus attempted to dispense with both radical utopianism of the Godwin sort and Adam Smith's sense of improvement in his irrefutable logic of population always trumping progress.

Malthus was *the* scientific theorist of anti-radicalism, opposing not simply the French Revolution – 'the fermentation of disgusting passions' as he put it – but also the very idea of a *right to subsistence* (and by extension a Painite theory of natural rights). Malthus had no difficulty in accepting the incursions on civil liberties in the name of social order – he would rather submit to 'great oppression than give the slightest countenance to popular tumult' – and by the same token rejected the right to subsistence as a 'delusive argument'. It was only fear of insecurity – driven by population growth – which could act as a spur to industry, indeed to progress. What troubled Malthus was precisely that the poor laws were 'a kind of systematic and certain relief on which the poor could confidently depend' (it was a sort of right). And this confidence carried two enormous burdens: first, it was antithetical to the spur of fear and insecurity (of the disciplining effects of the labour market), and second it encouraged the poor to stand up for the subversive notion of the right to food. The renewal of the Poor Law debate between 1830 and 1834 was in this sense about a simmering crisis of social control and class hegemony. Malthus' achievement was to construct a discourse of poverty which challenged the radical Painites and their theories of rights and to this extent reflects both a voice of the status quo (the structures of society will always remain unchanged he said) and a fear of the enfranchisement of the popular classes (of social disorder from below). Fear, said Adam Smith, was typically 'the wretched instrument of government'; for Malthus state oppression and violence were always a necessary source of order. The 'heart of flint' that Godwin saw in Malthus was perhaps less about population as such than Malthus' reluctance to admit – as Condorcet and Godwin foresaw – that there were no natural laws as such but rather a process of perpetual political consultation including consultation of people without political power.

Malthus, who has bequeathed such an important legacy to modern understanding of how we think about 'the population problem', must in other words be located on the larger canvas of competing ideas about nature and culture – discourses we might say – as they emerged in the context of industrial capitalism and liberal government. Glacken in general was less attentive to these sorts of historical semantics and to the traffic between particular meanings of nature and culture at the very birth of industrial modernity, and modern government. Glacken adhered to a 'core sense' of words (*contra* Williams) rather than the idea that semantics is a dynamic negotiation in the context of situated speech-acts. He also suffers (by default rather than design) from the search for an 'originary moment', a sort of Adamic urge that is really part of the logic of Whig historiography in which the past is plundered for precursors and founding charters that ratify the present. And not least Glacken's approach is overly textual. Language is the practical consciousness of *homo sapiens* and ideas are only a material force in the context of lived, embodied activity.

These three tensions in *Traces* point to another, perhaps more profound and deep, though related to each at the level of epistemology. Not only is he less attentive to the political economy of discourses of nature and culture, but he accepts that his three tropes of nature/culture frequently rest on a sort of *opposition*, not simply a distinction. They are binaries in which culture is nature's other. Glacken devoted much attention of course to Descartes and the 'ideal of a purposive control over nature through applied science' (1967: 427), and to the debates surrounding what Whitehead saw as the dreary mechanical universe of the closing years of the seventeenth century. But there is a larger epistemological question here – a product of the Enlightenment and the Cartesian and Kantian 'analytical mindset' (Castree 2003: 168) – of the proliferation of a number of important dualisms: human/non-human, representation/reality, mind/world and so on. This is not the place to explore the relations between the world of empire, commerce and science and how the very idea of the West emerges in opposition to remote places populated by people without culture (see Wolf 1977; Hall 1992). What matters is how pernicious this binary Cartesian epistemology became in the modern epoch. The Enlightenment understanding of scientific knowledge rested on several epistemological and ontological foundations including (i) that the physical world exists independent of human knowledge and judgement, (ii) that objects can be observed directly in a way that knowledge can be confirmed by reference to the real world and (iii) that truth refers to the correspondence between our ideas and nature and how it actually is (see Demeritt 2001). In this wider sense, Descartes' *cogito ergo sum* unleashed a world of forces and energies pushing and pulling bodies in a Newtonian world of attraction and repulsion in mechanical equilibrium. Implicit in this world-view was a 'social physics', the world as a billiardball table – what Gregory Bateson (1972) called an epistemology of linear causation or force or power. The Cartesian revolution involved two major impositions. First it imposed an ontological status upon entities (substance) as opposed to relationships (that is to say

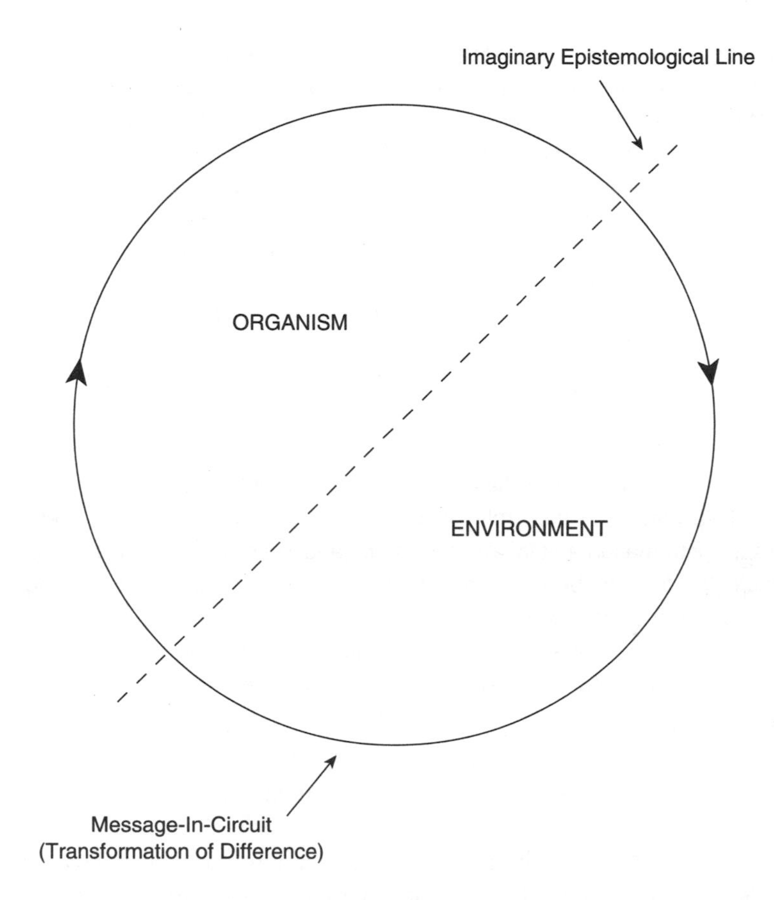

OPPOSITIONAL RELATIONS

ORGANISM	ENVIRONMENT
Man	Nature
Self	Other
We	Them
Ego	Superego (or Id)
Man	Woman
Reason	Emotion
Mind	Body
White	Non-white
'Civilized'	'Primitive'
Capital	Labour

Source: A. Wilden, 1972, p. 221

FIGURE 8.1 Nature/Culture: binaries and the epistemological Line

energy, matter, people, ideas and so on became things). And second, it imposed, as all epistemologies must, a line in which a logic of either/or (rather than both/and) predominated. As Anthony Wilden put it:

> Epistemology is a question of where you draw the line … the line between 'organism' (*or culture*) and 'environment' (*or nature*) by our conventional model of reality is such a line, and like all lines it is a fiction. Unfortunately we think it is real. (1972: 219; emphasis added)

Wilden, using the work of Gregory Bateson, laid out the 'epistemological error' as he saw it of dividing the world into these oppositional entities (Figure 8.1). The drawing of the lines generated a series of imaginary distinctions (that are taken as real) all of which rest upon a robust logic of cause, input–output, linearity and so on. For the likes of Bateson and Wilden, the epistemological alternative is what he calls a 'message in circuit', or 'circular reason':

> Nor is the unit of survival 'in' entities – 'in' the organism or 'in' the environment – the unit of survival is in their relationship, which is nowhere. It is nowhere because it is information and information is first and foremost difference … the epistemological line has been typically drawn in the wrong place … its placing corresponds paradoxically with a real, material locus of difference. The error is to turn the distinction between the loci of symbiotic and non-exploitative relationships into an Imaginary opposition of entities. (Wilden 1972: 222)

> The whole thinking that goes along with 'input' and 'output' is monstrously bad. It draws the line across a systemic structure … and it's me against the universe at once, the moment you draw the line. (Bateson 1974: 28)

Two things can be said about these binaries, and nature/culture dualism in particular. The attempt to think about a both/and epistemology did not of course begin with philosophers, cyberneticians, communications theorists and linguists in the 1960s. From the very birth of the Cartesian revolution itself there was always a counter-community (one thinks of Pacsal, Hegel, Leibnitz, Spinoza) that spoke the language of relations and relational worldviews (see Harvey 1996; Merchant 1980). Indeed, the history of pre-modern Europe and antiquity traced by Glacken has numerous examples of such relational thinking about the world (Aristotle being the most famous). And second, the binaries can, and have been, enormously generative and productive. In human geography, for example, one can see, to take two simple examples, the nature/culture binary at work in natural hazards and cultural ecology. The latter precisely examines the relations between nature and society by exploring the ways in which cultural forms could be adaptive with respect to the environment. Geographers and anthropologists could locate in ritual, or views of the universe, or local farming practices, an adaptive 'logic' or structure at work, isomorphic with the adaptive structures of all living systems: that is to say, culture has adaptive consequences in nature (Grossman 1984; Nietschmann 1972; Waddell 1975). Roy Rappaport's brilliant monograph *Pigs for the Ancestors* (1968), shows how a pig-slaughtering ritual in upland

New Guinea functioned as a thermostat to confine human populations within a sort of equilibrium with their fragile tropical rainforest. The 'cognized' model of the Tsembaga Maring elicited behaviour adequate to the 'operational model' of ecological science. A vast amount of work on ethno-botany and indigenous farming practice also explored the ways in which local forms of classification and agronomic practice were 'really' adaptive for the environment. Natural hazards (Burton et al. 1978) explored the question of human perception of and response to various 'environmental' perturbations and threats (drought, earthquake, tornado and so on). These behavioural and cognitive models were precisely based on a separation of nature and culture; human perception of the environment. Naturally there could be reactions – human responses could make environmental consequences better or worse. Even in influential contemporary social theoretical iterations – for example the work of Anthony Giddens (1991) – a recognition of the mediated character of nature is in the final instance superseded by the conventional binary categories of 'the transformation of nature' and 'ecological threats'. The epistemological line that Wilden describes remained, in sum, largely unexamined. As Harvey put it, organisms of any sort (human or non-human) work with evolutionary possibilities in dynamic and interactive ways and to grasp this fact requires that 'the artificial break between "society" and "nature" must be eroded, rendered porous and eventually dissolved' (1996: 192). But how exactly?

Evidently, these binaries can be, and have been, prodigiously generative (a large part of what passes as science is predicated on such epistemologies). All study involves some sort of line-drawing, and there are only a certain number of loci through which to draw it. One might ask, of course, why certain sorts of lines – why binary oppositions of the nature/culture sort – have such endurance, or why they became hegemonic? Bruno Latour pointedly observed that 'dualism may be a poor solution but it provided 99% of the social sciences' critical repertoire' (1991: 54). In his account it was the history of science that upset the applecart. In as much as historians of science opened up the black box of scientific knowledge production, the broader question of epistemic authority is an enormously complex topic that would take us deep into the trenches of the political economy of capitalism, the political economy of ignorance, the rise of the West and the Rest and much else (see Povey 1998). At its base, however, is the vexed question of what Jassanoff (1990) calls 'co-production': the simultaneous production of knowledge and social order. Here the framing of science and its problems, the creation and policing of boundaries, and the purification of scientific and other institutional conventions are all key to the robustness of dualistic thinking (see Forsyth 2003).

But these binaries are vastly generative workhorses, albeit with a cost. Gregory Bateson put it this way:

> Erroneous premises in fact work ... . On the other hand the premises work only up to a certain limit, and at some stage or under certain circumstances,

> if you are carrying serious epistemological errors, you will find that they do not work any more. At this point, you discover to your horror that it is exceedingly difficult to get rid of the error, that it's sticky ... the falsification gets around; and each thing you wipe it off on gets sticky ... (1972: 479)

What, then, are the circumstances under which these premises do not hold? Again the question is complex, having to do with the destabilization of normal science, the changing politics of institutionalized science, the 'anomalies' thrown up by the real world, and so on. But surely one reason why the epistemological question surrounding culture and nature has been subject to such critical scrutiny in the contemporary epoch turns on the material world of late modernity itself. Glacken did not explore the late twentieth and early twenty-first centuries of course, but the very world we now inhabit surely makes the binaries rather dubious. Glacken would have assuredly pointed to the fact that the relations between nature and culture in the late modern era duplicate older historical themes (the distant riders once again). The desire to *return to nature* (deep ecology, Earth First! and rusticators of various stripes, many variants of which, in search of an authentic *Volk*, come perilously close to the dark romanticism of 'blood and soil') surely has a long pedigree. The *return of nature* (global warming, the return of neo-catastrophism in the *fin de siècle* earth sciences, and the apocalyptic thinking that sees nature striking back in the form of Ebola or SARS) and the *end of nature* (the transcendence of limits promised by human genomics, the world of the clone and of artificiality, or the Benjaminian dream of cornucopian 'green' capitalism) are dotted throughout *Traces*'s long march through the Western tradition. But Glacken's 'horsemen in the dust' sensibility obscures a larger point powerfully articulated by Seigfreid Giedion in *Mechanization Takes Command* (1948: 6), namely: 'What happens when mechanization encounters organic substance?' This provocation has a particular saliency in the modern world. And the answers so often rub against the grain of what we take for granted, not readily fitting into the world of binaries. In Latour's language (1991) the profusion of entities like the ozone hole or genetically modified foods are 'hybrids'; the divide between the artificial and the natural is breached, and with it the two 'practical ensembles' (as Latour calls them) of modernity: namely that nature has an independent objective existence, and society is humanly constructed. Hybrids, however, were largely politically visible. But they have proliferated to the point where they can no longer be ignored. Let me sketch four salient examples.

Plants, birds, resources, zoos: modern hybrids

First, Michael Pollan's wildly popular book the *Botany of Desire* (2001) has shown in that book how botanical impulses shape some of our most abiding aesthetic, psychological and social desires. What he offers is a particular reckoning of the co-evolution of humans and the environments of which they are part, by honing in on a quartet of life forms, actually four familiar plants (the tulip, the apple, the cannabis and the potato), and four corresponding desires (beauty, sweetness, intoxication and control). Rather than accepting these plants as 'domesticated species', he views them as repositories of cultural (as well as natural) information; genes are, he says, 'archives of cultural as well as natural information' (2001: xvii). The tulip contains something of the aesthetics of the Ottoman Turk, the Russett Burbank potato of 'our taste for long, perfectly golden French fries' (2001: xvii). What is on offer is a natural history of human desires : human ideas and preferences and values end up as natural facts. Pollan in fact pushes further, trying to sustain an argument that such plants have *domesticated us*. As an antidote to what he calls the overestimation of our own agency in nature ('human invented agriculture', 'the human domination of nature' and so on). *The Botany of Desire* suggests that 'it makes just as much sense to think of agriculture as something the grasses did to people' (2001: xxi). Minimally, the agency of human is reversed in some way that makes culture and nature a dialectical unity.

Second, the chicken. We are living in, if I may quote none other than *The Economist*, the age of the chicken: 'From politics to the economy, from corporate lift-offs to consumer life styles, the late twentieth century is turning out to be the age of the chicken' (1997: 45). At the heart of this phenomenon is the *radical* mechanization and transformation of the chicken industry whether measured in terms of per capita consumption, labour productivity, gross output, or the 'productivity' (indeed the biology) of the chicken itself (days to maturity, average live weight and so on). It is all rather remarkable (see Table 8.1). Advances in breeding, disease control, nutrition, housing and processing have conferred upon US integrators some of the lowest production costs in the world. Since 1940, the industry's feed conversion rate (pounds per feed per pound of live broiler) has declined precipitously from 4.30 to 1.75. The average broiler liveweight has increased from 2.89 lb in 1934 to 5.02 in 2001 and the maturation period – the time required for a bird to reach market weight – has plummeted from over 95 days to less than 45. The result is what was called in the 1940s, the 'perfect broiler'. Avian science has now permitted the mind-boggling rates at which the birds add weight (almost 5lb in as many weeks!). The doubling of the average live bird weight in the past 50 years has seen, over the same period, the labour input in broiler meat production decline from 5.1 hours per 100 lb in 1945 to less than 8 minutes in 2000! The broiler is the product of a massive R&D campaign increasingly driven by genetic engineering (the chicken genome is almost fully mapped). Genetic engineering breeding, high nutritional growth regimes and disease regimens has made for a sort of steroidal chicken – grotesquely disfigured from its domestic progenitors – whose physiology is designed for the machine and the market. In order to maximize the production of breast meat and to facilitate mechanized slaughter and cutting, radical changes in the skeletal form of the chicken have been required. The broiler is a sort of Frankenstinian product of biotechnical mechanization (the nutrition–feed–health–confinement complex), consumer preference and the rigours of competitive capitalism. The chicken is, in short, a cyborg: part machine, part bird, part culture, part nature, but in an inseparable, one might say *organic* sense.

Third, petroleum as a 'natural resource'. It is sometimes said petroleum is the economic bedrock of our hydro-carbon civilization. The fuel of modernity, oil is an archetypal global commodity, the repository of unimaginable wealth ('black gold') and part of the largest business on earth, To be an 'oil nation' is to have discovered El Dorado in your backyard. It squirts 'obligingly up into the air' noted the great Polish journalist Ryzsard Kapucinski and 'falls back down to earth as a rustling shower of money' (1982: 45). But more than anything else, petroleum is a sort of lie: it reveals the profound mystification, the paradoxes and contradictoriness that surround natural resources in our modern world (Coronil 1997). What exactly, then, is natural about petroleum? Well, it is naturally oily. It is a flammable liquid that occurs as a product of geo-physical and biological processes of great historical depth. A by-product of pre-human geological history, oil is deposited in subterranean formations and consisting principally of a mixture of hydrocarbons with traces of nitrogenous and sulphurous compounds. In practice, of course, the composition of what passes as petroleum varies quite considerably, as one might anticipate in view of the heterogeneous circumstances associated with its 600 million year history of sedimentation and organic decomposition. Oil's natural properties, one might say, are unstable and variegated.

But there is another realm in which all natural resources must operate, namely the social imaginary: in other words, how oil is rooted in the imagination of people living in the specific historical and social circumstances of its use and deployment. Oil as a natural resource carries it own mythos, shaped by place and time. I currently inhabit a petroleum age in which, from the vantage point of the oil-importing North Atlantic economies, oil stands in a specific relationship to the mosque and the Arab world. For oil-producing states, petroleum provides the idiom for nation-building and the financial wherewithal for modern development (think, for example, of the petrolic-ambition of a great modernizer like Shah Palavi in Iran). Oil is inextricably bound up with unimaginable personal power (Rockefeller, Nobel, Rothschild, the Sultan of Brunei), untrammelled corporate hegemony ('the Seven Sisters'), and a history of spectacular imperial violence and war. Did not the long and dark tentacles of oil appear in the catastrophic demise of the Twin Towers of the World Trade Centre in New York? Was not Osama bin Laden a product of oil as much as of Wahabbi Islam? Was not the Ayatollah Khomeini's revolution in 1979 fuelled by oil-inspired resentments and grievances? Need I mention Enron? Oil and Islam, war and violence, corruption and power, wealth and spectacle, scarcity and crisis are, in our times, seemingly all of a piece.

TABLE 8.1 **The chicken revolution: US broiler production and consumption 1934–2000**

Year	US production (million heads)	Increase over previous period (per cent)	Maturation (days)	Feed conversion rate*	Labour input (hours)**	Price per pound (dollars) live	Price per pound (dollars) r-t-c[1]	Average live weight (pounds)	Per capita consumption (pounds) r-t-c
1934	34	–	112	4.30	5.1	0.193	–	2.84	0.7
1940	143	321	–	–	–	0.173	–	2.89	2.0
1945	366	156	–	–	–	0.295	–	3.03	5.0
1950	631	72	–	–	–	0.274	–	3.08	8.7
1955	1,092	73	–	–	–	0.252	–	3.07	13.8
1960	1,795	63	–	–	–	0.169	–	3.35	23.6
1965	2,334	30	–	–	–	0.150[2]	26.5	3.48	29.9
1970	2,987	28	–	–	–	0.135	26.4	3.62	36.8
1975	2,950	–1	–	–	–	0.262	45.1	3.76	36.8
1980	3,963	34	–	–	–	0.279	49.1	3.93	46.6
1985	4,470	13	–	–	–	0.302	50.8[3]	4.19	52.0
1990	5,864	31	–	–	–	0.324	54.8	4.37	61.0
1994	7,018	20	–	–	–	0.350	55.7	4.63	69.9
2000	8,262	18	45	1.75	0.12	0.336	58.14	5.00	89.6

*Pounds of feed per pound of live broiler
**Hours per 100 lbs of broiler meat

Notes:
[1]ready-to-cook.
[2]live price changed to live weight equivalent price in mid-1960s.
[3]9-city wholesale price changed to 12-city composite wholesale price in 1983.

Source: Poultry Tribune, September 1995, p. 1. *Watt Poultry Journal* (2000–2001), *NASS Poultry Quarterly*, various issues.

Petroleum is the quintessential modern natural resource. It is present in and produced by nature, and a material source of wealth that occurs in a natural state. But this is a contradictory and non-sensible claim on its face. Oil is natural insofar as it resides in its Jurassic bedrock. But it is not immediately accessible or useful; it presupposes human knowledge and practice (drilling, exploring, refining). Oil's wealth is not conferred solely by natural process but rests upon an appraisal – a state of knowledge and practice – that is social, technological and historical. Petroleum is profoundly *of* nature – it is typically subterranean and has peculiar biophysical properties. And yet its naturalism is expressed and understood in quite determinate ways; how differently would the first-century BC Chinese bureaucrat and the twentieth-century hard-rock geologist have described petroleum's natural properties! Petroleum's 'resourcefulness' is not natural at all. Its expressive form as wealth – the defining property of a resource – presupposes acts of transformation, distribution and use which, incidentally, was very different for sixteenth-century North American Indians than for a twenty-first-century Louisiana petrochemical industry. Petroleum as a natural resource rests, then, on particular meanings of natural (for example theories of biophysics), and particular renderings of resource (for example theories of wealth predicated on scarcity and natural limits).

Finally, the zoo. John Berger famously observed that animals constitute a living monument to their own demise: 'everywhere' he says 'animals disappear' (1980: 24). In posing the question 'Why look at animals?', Berger sought to show that there is a central paradox residing in our experience of the zoo and its inhabitants. It is a source of popular appeal and considerable attraction – 'millions visit the zoos each year out of a curiosity which was both so large, so vague and so personal that it is hard to express' (p. 20) – and yet in the zoo 'the view is always wrong ... like an image out of focus' (p. 21). Inevitably zoos disappoint. The excitement of the wild is replaced by alienation, lethargy, isolation, incarceration and boredom. The spark which has historically linked man and animal has been extinguished. At the heart of the zoo's paradoxical status is a sort of double alienation. On the one hand the zoo is a sort of prison – a space of confinement and a site of enforced marginalization like the penitentiary or the concentration camp. And on the other, it cannot subvert the awful reality that the animals, from whatever vantage point they are viewed, are 'rendered absolutely marginal' (p. 22).

The inescapable fact is that the zoo recapitulates the relations between humans and animals, between nature and modernity (see Rothfels 2002). It *demonstrates*, as Berger says, a basic ecological fact of loss and exclusion – the disappearance and extinction of animals – through an act of incarceration. Berger was at pains to connect the zoo as a monument to loss, as a space of confinement, to a human crisis, specifically to the disposal, or perhaps more appropriately the enclosure, of the peasantry. In quick succession, the emergence of the zoo – that is to say, the marginalization of the animal – is followed by the marginalization of a class – the peasant – for whom familiarity with, and a wise understanding of, animals is a distinguishing trait.

Berger offers a productive way of thinking about nature and culture. To put the matter starkly, one might say that the relation between animals (or more properly, nature) and modernity can be construed as a gigantic act of *enclosure* – necessitating, of course, loss and displacement – which contains a double-movement. The first, of which the zoo is the exemplary modern instance, is to accomplish what Stephen Greenblatt in *Marvelous Possessions* calls 'the assimilation of the other' (1991: 3). Greenblatt suggests that the zoo, for example, is part of the 'reproduction and circulation of mimetic capital' (p. 6). The zoo is an example of what he calls a cultural storehouse, part of the proliferation and circulation of representations which become for Greenblatt:

> a set of images and image-making devices that are accumulated, 'banked' as it were, in books, archives, collections, cultural storehouses, until such time as new representations are called upon to generate new representations. (1991: 6)

This mimetic quality of capitalism suggests that the representations – of animals, of ecosystems, of nature – are social relations of production. The representation is both the product of the social relations of capitalism and is a social relation itself 'linked to the group understandings, status hierarchies, resistances and conflicts that exist in other spheres of the culture in which it circulates' (Greenblatt 1991: 6). The zoo is, then, product and producer. The zoos places animals in captivity ('preservation', 'salvage ecology') as a response to – the product of – the devastating ecological consequences of modernization. Equally, the zoo culture also serves as an institutionalized means, a scientific means no less, to represent animals in quite specific ways and generate culturally mimetic portrayals of itself.

These four vignettes suggest, in quite different registers, that the line between nature and culture has been irrevocably ruptured. Perhaps this was always the case, but science, technology, forms of representation, the process of commodification, the spectacular culture of capitalism, have all somehow conspired to produce a mixing, or a hybridization, for which the old binaries make even less sense. To the extent this represents a sort of challenge for the academic disciplines, there is, and has been, no one singular response. On the one side is a hardening of the binaries in the face of challenge (seen, for example, in the Culture Wars – see Eagleton 2000). In others, the binary is breached by a sort of reduction or essentialism, that is to say a collapsing of the distinction. Nature is simply folded into culture (the *culturalism* of the radical forms of social constructionism) or culture into nature (the *naturalism* of the Bell Curve and the 'renaturalization' of inequality). The sort of essentialism by which, for example, gender and nature are organically conflated in certain versions of Western feminism is a powerful reminder of how the politics of such reduction can produce deeply problematic politics of the blood and soil variety (see Shiva 1988; and Ferry 1995 on ecofascism). But there are also efforts within geography (and outside) to take up Bateson's challenge of an epistemology of both/and. And it is to these efforts that I now briefly turn.[5]

The intellectual field of what we might call non-binary – what in the French tradition is called 'non-centred' (Whiteside 2002) – perspectives on nature/culture is extremely varied, and any mapping would have to encompass lines of thinking that span the human and bio-physical sciences. The genealogy, moreover, has substantial historical depth in the modern epoch, often operating under the sign of relational thinking. The philosophy of internal relations can be traced to the Greek philosopher Parmenides, and in the modern period to the work of Spinoza. But for our purposes it is Marx – despite the Promethean tendencies with which he is customarily tarred – who provides the most elaborate effort to what we could call the 'dialectics of internal relations' (Ollman 1972). It was Marx after all of whom Pareto said that his use of words was peculiar; in them one can see both bats and mice. When we reflect on nature, said Engels, the 'first picture presented to us is of an endless maze of relations and interactions in which nothing remains what, where and as it was but everything moves and changes' (1978: 26–7). This type of reasoning can also be discovered in pre-capitalist settings. Reichel-Dolmatoff's extraordinary ethnography *Amazonian Cosmos* (1972) charts an indigenous Amazonian 'theory' of the universe and of relations between culture and nature that looks suspiciously like a sophisticated relational view of the world. Taussig's (1980) account of devil cults in the mines and plantations of Bolivia and Colombia emerges from a 'logic' of pre-capitalist reflection. Products may have fetishistic or life-like qualities, but they emerge from 'the sense of organic unity between persons and their products' (1980: 37); nature cannot resist man and man cannot resist nature.

To make this field somewhat manageable I have chosen to identify three *topoi*, largely within human geography but drawing upon wider theoretical frames of reference, which overlap in all manner of ways but which have at their centre a discrete way of thinking in non-binary ways about nature and culture[6].

Dialectics/production

The Marxist tradition is variegated and heterogeneous, and there is no easy way to navigate through the sea of contradictory opinion on whether Marx himself considered nature to matter and how (see Castree 2003). What is less controversial is the extent to which Marx sees the relations between humans (as culture bearers and producers) and nature as constituted through labour and the labour process. The organization of labour and its means of work transforms bio-physical nature – in the act of appropriation – and in turn changes human nature. On occasion Marx referred to the nature/culture relation as a 'metabolism', which is to say it is relational and dialectical. All things in Marx were seen relationally – as a process – and this was no less the case for nature, which exhibited dialectical and internally related qualities: the transformation of quantity to quality, mutual penetration of opposites and the transformation into each other at the limit case, development through contradiction, and spiral forms of development (Ollman 1972: 55). Writing in the midst of the ravages of nineteenth-century industrial capitalism, Marx saw that the advent of industrial modernity not only heralded new forms of what he called social labour (factories, new divisions of labour and so on), but also how nature itself was refashioned. What was at stake was the *production* of nature. Neil Smith formalized this line of thinking in his book *Uneven Development*.

> In its constant drive to accumulate ... capital transforms the shape of the entire world. No god-given stone is left unturned, no original relation with nature is unaltered, no living thing unaffected. Uneven development is the concrete process and pattern of the production of nature under capitalism. (Smith 1990: xiv)

The consequences of seeing nature as produced is, of course, to explore how such productions also serve ideological ends. Ideologies of nature or ideologies of science are constitutive of nature under capitalism, obscuring yet ultimately serving particular class interests and dominated finally by the irreducible logic of capitalist accumulation.

The revolutionary capacity of capital to produce a 'second nature' out of the pristine first nature was of course foreseen in the realm of agriculture by Kautsky in his great book *The Agrarian Question* (1899) at the end of the nineteenth century. Capital was taking hold of the point of production, smashing the old relations of production but holding out the technological promise of 'making bread from stone'. The danger in this line of thinking is a strong form determination dressed up as relational thinking. In this sense the debate initiated by Mann and Dickinson (1978) sought to reinsert the physical and biological properties of the living system as a source of resistance to the logic of commodification – a barrier to accumulation. In an echo of Polanyi's (1967) work on fictitious commodities – those production conditions like nature that could not be produced as commodities – so-called ecological Marxists sought to unravel the contradictions between economic and biological logics. The disjunction between biological time and productive time, and the effects of biological degradation on long-term profitability, drove home the point that Marx's emphasis on the labour process must recognize nature limits (Benton 1989; Henderson 1999). Nature was in some way constitutive of the particular forms of accumulation (perhaps say the persistence of peasant production in the face of agribusiness as the former assumed the biological risks of the latter).

The capacity to produce or make nature through, for example, genetic engineering – the idea of a design of nature harkening back to Glacken but in a very different register – takes on a new meaning in the late twentieth century. The patenting of life forms, the creation of genetically modified crops and the prospects of cloning within and across species represent a radical break, opening up the possibility for a sort of 'real subsumption' of nature as accumulation operates through nature, or more properly, rather, nature is manipulated and made in a way that it becomes a force of production (see Goodman 2001; Schurmann et al. 2003). But how is the agency of nature to be retained? It is here that the addition of a dialectical twist to the classic expressions of the production of second nature under capitalism – what Castree calls 'coupling and conjoining' (2003: 174) – is so productive. For Harvey (1996) the nature/culture binary is seen dialectically as a socio-ecological project, a 'created ecosystem' which instantiates and reflects the social systems that gave rise to it. Human beings, like other organisms, are active subjects transforming nature according to its laws and are always adapting to the ecosystems they themselves construct (1996: 186). Socio-environmental life is an indissoluble process of 'multiple moments' and 'material permanences', and the study of them is part of this process (hence situated and provisional). More formally, Harvey sees the socio-environmental projects as part of an 'evolutionary process' constituted by four moments:

1 Competition (the production of hierarchy and homogeneity).
2 Adaptation/diversification (the production of diversity).
3 Collaboration/cooperation (the production of social forms).
4 Environmental transformation (the production of nature).

Each of these categories is relational, and each internalizes the effects of the others. As he makes clear, there are different modes of competition, adaptation, cooperation and transformation and the way in which they are linked (internalized) is heterogeneous and therefore a source of tension and conflict. In his words:

> I am proposing a way of depicting the fundamental physical and biological conditions and processes at work through all social, cultural and economic projects to create a tangible historical geography and to do it in a way as not to render those physical and biological elements as a banal and passive background … and to specify the conditions and processes in such a way as to understand the possibilities for collective human agency in negotiating through these fundamental elements to generate significantly diverse outcomes. (1996: 192)

In this way the dialectical vision ends up on a terrain quite close to that articulated by the biologists Bateson (1972) and especially Lewontin (1982: 3), for whom organisms are 'subjects transforming nature according to its laws'. Harvey's account of what he calls the 'value form under capitalism' and how it envelops nature discursively and physically, and Swyngedouw's (1999) account of waterscapes in Spain and Ecuador as a global/local inter-waving of people, things and artefacts under the aegis of capital are two suggestive studies of the dialectical approach at work.[7]

Construction/representation

What might it mean to see nature/culture as a unitary social construction (to claim that nature is part of culture and always mediated by it (Wilson 1991)? To pose the question in this way presupposes at least minimally a working definition of what is involved in social construction. There are several different senses in which the term can be deployed: as refutation, as critique of Enlightenment knowledge presumptions, as representation, as rendering visible (see Demeritt 2001; Gregory 2001). Underlying this heterogeneity of what might be seen as stronger or weaker forms of construction is a twofold concern with challenging authorized and unquestioned knowledge (especially science) and seeing nature not as a given or taken-for-granted. Put differently, construction starts from the question: why do we know what we know (and relatedly what we don't know, that is, the political economy of ignorance)? Why are some forms of knowing more influential and valid than others (what are the conventions of legitimation and authorization)? The charge is to explore the contours of hegemony, of ruling ideas and practices, while always attentive to the situatedness of knowledge, to the institutional context of its production and reproduction of knowledge, and to the contradictions and tensions of securing legitimacy.

Some of the most generative work on social construction has explored nature/culture through the tropics and tropicality, often in relation to empire and imperialism (Arnold 1996; Stepan 2001). The tropics can be seen as a sort of geographical imaginary that contains within it the traces of colonial domination and normalization (Gregory 2001). In the nineteenth century two articulations of tropicality predominated. One was an edenic narrative of tropical exuberance and fertility and the other of pathology and monstrous excrescence. As the century wore on, the two became entangled. Stepan's exemplary book *Picturing Tropical Nature* (2001) examines how tropical nature became a landscape through the works of illustrators, artists, explorers and the likes of such scientific figures as Wallace and Humboldt. The tropicalization of nature replicates many of the broad themes of pathology, enframing, domination and projection recounted by Arnold and Gregory, but as Stepan shows, the relation between production/construction and consumption/interpretation is always tense and contingent. Louis Agassiz's vision of Brazil as a space of disease and racial miscegenation – embodied in his influential photographs – were in turn appropriated and deployed by Brazilian intellectuals to create a mythology of Brazilian 'racial democracy'. Even so, as Gregory shows, the tensions between representations of the tropics as a way of securing the superiority of the domestic over the exotic (the West over the Rest) was always riven by the prospects of pathologies and transgression: 'the hideous threat that ... monstrosities would break out from the place that the colonial discourse had assigned to them' (2001: 98).

It is no surprise that more recently Amazonia has figured centrally in regard to how certain sorts of postcolonial natures have been rendered visible. Tropical nature, says Stepan, is a way of writing but it is also a way of storytelling for multiple peoples and multiple constituencies inside and beyond the space of Amazonia. Slater's book *Entangled Edens* (2002) tackles the question of multiple narratives head-on, exploring these tangled edenic visions and the charms and their complex histories all the while sensitive to the fact that Amazonia is simultaneously place and icon, natural creation and human invention. Hugh Raffles' *In Amazonia* (2002) takes the reader through the long march of Amazonia's many natures – its natural history. It is an ambitious, and I think ultimately flawed, attempt to, as he puts it, write 'the present as a condensation of multiple natures and their differences' (2001: 7).

Another discursive entry point into the construction of the tropics interrogates the field of contemporary science and policy rather than colonial history. The ground here is provided by 'conventional models', as Leach (1997) calls them – a co-production of environmental knowledge and political practice – and by extension, why and how particular regulatory institutions and practices become hegemonic and are then subsequently contested. Latour (1987) talks of how particular problems are 'black boxed' precisely as a way of flagging how conventions are rendered objective and secure. Neumann's excellent book *Imposing Wilderness* (1999), on the creation of the Arusha national park in Tanzania and the ideas of landscape and nature which lay behind state appropriations of land from local peasant communities, is an especially compelling illustration of how cultural and historical representations of nature intersect with colonial and post-colonial rule. Leach and Mearns' (1996) reinterpretation of the West African forest–savanna mosaic is a careful deconstruction of a conventional model in which historical studies coupled with detailed local analysis of agro-ecology confirm what the new 'non-equilibrium' ecology posits, namely that climax models of ecological stasis are unhelpful. These static models nonetheless enter into administrative practice (colonial and post-colonial), reinforcing the idea of Guinée's forest cover as 'relic' (which Mearns and Leach see as the basis for driving 'repressive policies designed to reform local land use practice') rather than seeing it as the outcome of intentional local management practice. Swift (1997) has documented how the assumptions about desertification not only rest on remarkably sparse evidence but on questionable models of the dynamics of semi-arid rangelands – their resiliency and stability in other words – which are expressions of linear, cybernetic models of ecological structure and temporalities, and are attached to neo-Malthusian models of social change. The key here is that conventional wisdom is challenged as an embodied form of knowledge, and the challenge itself reflects a peculiar unity of local knowledge and practice with non-linear models of new ecology. Out of this emerges a concern with pluralism (at the level of truth claims), with democracy (to open up the practices of policy-making to other voices), and complexity/flexibility (of local conditions and historical dynamics).

One response to the question of authority and legitimacy turns to the culture of expertise and so-called 'epistemic communities'. Here the knowledge is Western science, and more properly the cosmopolitan scientist-expert-policy-maker. Peter Haas (1990) has argued in the context of understanding regional (European Union) and global (multilateral) conventions that the process of consensus building and collective action more generally is *knowledge-based* and *interpretive*. That is to say, international regulatory cooperation is fuelled by fundamental scientific uncertainty about the environment which ensures that governments seek out authoritative advisers (experts) who, to the extent they are part of epistemic communities, are more important to the political solution than the content of the ideas *per se*. On a much broader landscape, the expert is part of what Beck (1992) calls the 'risk society'. Discourses and representations of risk – typically through scientific and technocratic expertise – can only be grasped in relation to what Beck calls 'reflexive modernization' (the incorporation of environmental awareness, including new post-industrial threats, into the modern political project) and the collapse of the classic industrial 'safety state'. Politics increasingly becomes, then, a discursive struggle over who is to determine what represents harm, who arbitrates such risks and harms, what is to count as evidence and proof, and who is to adjudicate (see Hajer 1995; Harre et al. 1999). In this setting, as Hajer notes in his critique of ecological modernization, storylines are essential political devices to permit what he calls 'discursive closure' in the name of risk alleviation or mitigation. There is much that is too general and overarching in Beck's corpus, but the very idea of a risk society sharply poses the question of how technoscience and advanced capitalism have constructed a particular unified vision of nature/culture encompassed through a powerful institutional and discursive nexus of risk and insecurity. In this sense the construction of nature/culture must be grasped not only as discourses but as discursive practices that 'enter fully into the constitution of the world' (Gregory 2001: 85).

Hybridity/practice/calculation

Timothy Mitchell (2002) opens his monograph on Egyptian modernity with the provocative question 'Can the mosquito speak?' He is concerned with the post-1945 modern development of Egypt, and the bundle of plans and projects around the Aswan dam, irrigation, agricultural reform, health and modernization. This endeavour involved, in Mitchell's account, a concentration of knowledge as a centre of state calculation and expertise, a series of catastrophic failures and a techno-science intent on covering its failings and inadequacies. In all this, science was opposed to nature in a series of binaries, but this binary world could only emerge 'by engaging a series of other logics, forces and chemistries' (2002: 51) including the River Nile, the mosquito, the world addiction to sugar, and so on. In all the struggles, plans and politics of Aswan, the mosquito figured centrally. As he says, the mosquito was put to work. On his account it is not simply about nature's resistance or non-human forces working against development. As he puts it:

> What is called nature or the material world moves, like the plasmodium, in and out of human forms or occurs as arrangements, like the River Nile, that are social as well as natural, technical as well as material. The world out of which techno-politics emerged was an unresolved and prior combination of reason, force, imagination and resources. (2002: 52)

What matters is not binaries but 'hybrid agencies' (p. 53).

Mitchell's account of the mosquito as a material presence in the assemblage of forces in the production of Egyptian modernity recalls Callon's (1986) extraordinary account of the domestication of scallop production in St Breuc Bay. He treats the scallop as an agent, thereby breaching the protocol of causal asymmetry in the relations between nature and culture. The scientists, the scallop collectors and the scallops themselves are recruited into a particular network in order to support the particular explanation of the production of scallops under review. Both Mitchell and Callon inject a natural agency into the multi-layered and complex assemblages or networks of nature/culture and more ambitiously propose a model of 'extended translation' of conventional science (or techno-politics in Mitchell's lexicon) in which social framings of the world create hybrid creations of social and physical objects: quasi-objects in Latour's language, cyborgs in Haraway's (1992).[8] The nature/culture dualism in the face of such hybrid objects and multiple agencies can only be maintained through a purification process; a similar function in Mitchell's analysis is played by the omnipotent logic of capital. In both cases, what matters is an artefactual nature not captured by either nature or culture. The obverse holds, namely that the nature/culture dualism is a sort of enframing, a picturing, in which nature is staged, made available to us through multiple practices and performances – imaginaries in sort – each of which 'stands in for the objective, original and enduring structure of the natural world' (2000: 24).

There are two lines of theorizing that shape the emergence of the hybrid and act as a metaphor for non-centred views of nature/culture. The first emerges from the work of Michel Serres and Bruno Latour. Serres is a mathematician whose book *The Natural Contract* (1990) seeks to reconceptualize the natural world as a field of passive objects available for appropriation or destruction. A reciprocity with nature demands a natural contact in which we recognize that the founding concepts of science affect its topography. An understanding of the natural world should be informed by a sense of fluidity rather than order and contractual confederation rather than domination and mastery. Latour sees Serres as an anthropologist of science who provides the groundwork for a political ecology preoccupied with the modern problem of coping with unconstrained artifice. Latour diminishes the role of human agency and accentuates the role of 'actants' which may be people or bacteria (1991). The networks of humans and non-humans out of which scientific facts emerge are so entangled that the distinction between subjective agents and natural objects breaks down. The asymmetry between human and natural agency must be replaced with, to use Serres' language, a 'pragmatogony', an account of the reciprocal constitution of human subjects and the objects of the world. While the modern world is full of hybrids, the modern constitution denies them, and renders them unrepresentable. Rather than have nature represented (spoken for) by scientists, Latour argues for an assembly – a parliament of things – in which scientists, industrialists, urban planners and meteorologists and so on would debate and legislate on hybrids. It is the network – hence the moniker actor-network theory (ANT) – rather than the binary that captures modern nature.

The second genealogical thread derives from Foucault (see 2000) and the practices of biopower as a distinctively modern form of power. Government for Foucault referred famously to the 'conduct of conduct', a more or less calculated and rational set of ways of shaping conduct and of securing rule through a multiplicity of authorities and agencies in and outside of the state and at a variety of spatial levels. In contrast to forms of pastoral power of the Middle Ages from which a sense of sovereignty, was derived, Foucault charted an important historical shift, beginning in the sixteenth century, toward government as a right manner of disposing things 'so as to not lead to the common good ... but to an end that is convenient for each of the things governed' (2000: 211). The new practices of the state, as Mitchell Dean (1999: 16) says, shape human conduct by 'working through our desire, aspirations, interests and beliefs for definite but shifting ends'.

> On the contrary, in [the modern exercise of power], you will notice that the definition of government in no way refers to territory: one governs *things*. But what does this mean? I think this is not a matter of opposing things to men, but rather of showing that what government has to do with is not territory but, rather, a sort of complex composed of men and things. The things, in this sense, with which government is to be concerned are in fact men, but *men in their relations, their links, their imbrication with those things that are wealth, resources, means of subsistence, the territory with its specific qualities,*

> *climate, irrigation, fertility, and so on; men in their relation to those other things that are customs, habits, ways of acting and thinking and so on; and finally men in relation to those still other things that might be accidents and misfortunes such as famines, epidemics, death and so on ...* What counts is essentially this complex of men and things; property and territory are merely one of its variables. (Foucault 2000: 208–9; emphasis added)

It was Foucault's task to reveal the genealogy of government, and the origins and modern power, the fabrication of a modern identity. The conduct of conduct – governmentality – could be expressed as pastoral, disciplinary or as biopower. Modern governmentality was rendered distinctive by the specific forms in which the population and the economy was administered, and specifically by a deepening of the 'governmentalization of the state' (that is to say, how sovereignty comes to be articulated through the populations and the processes that constitute them). What was key for Foucault was not the displacement of one form of power by another, nor the historical substitution of feudal by modern governmentality, but the complex triangulation involved in sustaining many forms of power put to the purpose of security and regulation.

Governing, that is to say what authorities wanted to happen in relation to what problems and objectives through what tactics, can be assessed through the 'analytics of government' (Dean 1999; Rose 1999: 21), in other words, the processes by which we govern and are governed within different regimes, the conditions under which they emerge, operate and are transformed. Dean notes that there are four dimensions to government so construed. The first he calls *forms of visibility* (the picturing and constituting of objects). The second is the *techne of government* (through what means, mechanism, tactics and technologies authority is constituted and rule accomplished). Third, the *episteme of government* (what forms of thought, knowledge, expertise, calculation are employed in governing and how form is given to what is governable). And forth, *forms of identification* (the forming of subjects, selves, agents, actors, in short the production of governable subjects). Implicit in the notion of biopower and the analytics of government is the relation between nature as a realm of calculation, as a visible field, and particular forms of rule and subjectivity. Put differently, the sorts of binaries and dualisms associated with nature/culture – the modern constitution and its form of purification as Latour would have it – that are the very stuff of modern rule (biopower) in Foucauldian terms.

This confluence between hybrids (and the practical ensembles of modernity of nature/culture binaries) and what one might call green governmentality have been explored in a number of important ways. Whatmore's (2002) work on hybridity in human geography is deeply influenced by ANT and is put to the purpose of exploring the topologies of wildlife and the struggles over boundaries and enframing in struggles over plant genetic resources. Drayton (1996), from the vantage point of Kew Gardens and imperial science, and Koerner (1999) from the perspective of Linneaus and nation-building, have explored the ways in which nature and the analytics of governmentality move in tandem. But most powerfully, Braun's (2002) and Moore's (2005) works are the fullest account of a productive intersection of Latour and Foucault. Braun shows in the Canadian case how scientific framing through geology was central to the staging of the 'theatre of nature' after which land could be made available to political and economic calculation. Moore reveals land and conservation in Zimbabwe as an 'assemblage' of nature and culture.

Coda

> For it is hard to discover
> The winged vertebrates of prehistory
> Embedded in tablets of slate
> But if I see before me
> The nervature of past life
> In one image, I always think this has something to do with truth
>
> W.G. Sebald, *After Nature* (2003: 83)

Dialectics, construction and hybridity are all tropes for thinking about a non-centred culture/nature. They are clearly internally related in all sorts of ways, and in some cases totter on the brink of unintelligibility as they push the limits of non-linear thinking. Hybridity can sometimes degenerate into little more than a description of 'imbroglios' and networks that enmesh us, and seems inimitable to a weighting of social forces. Social construction can be annoyingly ambiguous on the point at which representations have efficacy in the material world. And all three unconvincingly theorize nature's agency in a way that its causal or actant powers are rendered weak and shadowy.

But these *topoi* are all distinctively modern in at least two senses and perhaps are best understood in terms of the contradictions that modernist senses of nature and culture have generated. First, they are all framed by the overwhelming contemporary reality of primitive accumulation – the process of dispossession – and enclosure, the ceaseless search for profit within new realms of the natural, always commodifying, privatizing and calculating in its wake (Harvey 2003). The fact that the DNA of some upland Papuan tribes has been patented is simply the most grotesque case in point. The second, is that within the circumference of the modern, there are dark forces at work within culture that spring from its 'troubled interface with nature' (Eagleton 2000: 110). Marx sees this darkness within the violence and conflict generated by the conditions under which the labour process appropriates nature. For Freud the traffic between the infant and her environment sows the seeds of destructive impulses and desires. For Nietzsche the dominion over nature involves what Eagleton calls 'a potentially catastrophic sovereignty over ourselves' (p. 110).

> Nature is not just the Other of culture. It is also a kind of inert weight within it, opening up an inner fracture which runs all the way through the human subject. We can wrest culture out of nature only by harnessing some of our own natural energies to the task; cultures are not in this sense built by purely cultural means. But these dominant energies then tend to set up a well-nigh unstoppable momentum which is far in excess of what culture needs to survive and which we can turn with equal aggressiveness upon ourselves. (Eagleton 2000: 110)

The culture/nature binary has obviously a natural history but the extent to which the alternatives outlined here, however undeveloped they may be, can avoid the sorts of aggressions and ruptures referred to by Eagleton is a matter of considerable conjecture.

FURTHER READING

Du Puis, Melanie (2002) *Nature's Perfect Food.* New York: New York University Press

Goodman, A., Heath, D. and Lindee, S. (eds) (2003) *Genetic Nature/Culture.* Berkeley, CA: University of California Press

Greenough, P. and Tsing, A. (eds) (2003) *Nature in the Global South.* Durham, NC: Duke University Press

Moore, D., Pandian A. and Kosek, J. (eds) (2003) *Race, Nature and the Politics of Difference.* Durham, NC: Duke University Press

Spufford, Francis (1997) *I May be Some Time: Ice and the English Imagination.* Picador: New York

Notes

1 Glacken's psychological health deteriorated in the latter years of his life, a condition that can be traced back to the Vietnam War and the protests on the Berkeley campus (when he served to his chagrin as Chair of the Department of Geography). His sequel to *Traces* was completed in the early 1980s but was not well received by the University of California Press, which distressed Glacken greatly. While it appears that Glacken destroyed the manuscript as his condition deteriorated in the 1980s, substantial portions of chapters have been located and are currently housed with the Glacken papers in the Bancroft Library on the Berkeley campus.

2 His book on population was the most widely read tract in the wake of Adam Smith's *Wealth of Nations.*

3 This is a criticism that I think holds for David Harvey's (1974) otherwise exemplary essay on Malthus, and Smith and Ricardo.

4 For example, his pessimism of the first essay is subsequently superseded by a more optimistic tenor in later editions which recognizes the role of moral restraint, 'irregular practices' (birth control) and education of the working classes as 'preventive checks', potentially unshackling the fetters of overpopulation. Likewise his writing on population is shaped by overpopulation (the Darwinian Malthus) but his *Principles of Political Economy* (published in 1820) turns on the problem of underconsumption and ineffective demand (the 'Keynesian' Malthus)! His theodicy of the first essay actually ran afoul of the Church, not least because he seemed to argue that natural laws superseded civilizational norms. And not least, while his harsh approach to the abolition of the Poor Law reeked of a high Tory elitism, he could also speak in the strongest terms for universal free education and civil liberty: as he put it 'No people can be much accustomed to form plans for the future who do not feel assured that their industrious exertions will be allowed free scope .... Civil liberty cannot be permanently secured without political liberty.'
5 For those interested in parallel efforts in other disciplines see Bohm (2002) for physics, Lewontin (1993, 2001) for biology, Luhmann (1991) for sociology.
6 Castree (2003) has provided a mapping of Marxist views of nature that covers similar ground. Readers should make use of his exemplary work (also see Castree and Braun 2001). My taxonomy combines some of his own categories (determination, articulation, conjunction, materialization) to provide for what I take to be more theoretical consistency. For a parallel effort to understand the traffic between culture and the economy see Sayer (1997).
7 Castree and MacMillan (2003: 210) see this dialectical approach as nothing more than a sleight of hand, substituting both/and for either/or while simply retaining a more 'permeable' divide. This is a curious claim since the very idea of full permeability across a dualism can be surely read as the sort of 'transcendence' they are looking for.
8 'Organisms are biological embodiments: as natural-technical entities, they are not preexisting plants, animals ... etc., with boundaries already and waiting the right kind of instrument to note them accurately. Biology is a discourse, not the living world itself. But humans are not the only actors in the construction of the entities of any scientific discourse: machines ... and other partners ... are active constructors ...' (1991: 298).

References

Anderson, K., Domosh, M., Pite, S. and Thrift, N. (eds), (2003) *Handbook of Cultural Geography*. London: Sage.
Bateson, G. (1972) *Steps to an Ecology of Mind*. New York: Ballantine Books.
Bateson, G. (1974) 'Scattered thoughts for a conference on broken power', *CoEvolution Quarterly* Winter: 26–7.
Beck, U. (1992) *The Risk Society*. London: Sage.
Benton, T. (1989) 'Marxism and natural limits', *New Left Review* 187: 51–81.
Berger, J. (1980) *About Looking*. New York: Pantheon.
Bohm, D. (2002) *The Implicate Order*. London: Routledge.

Braun, B. (2000) 'Producing vertical territory', *Ecumene*, 7: 7–46.
Braun, B. (2002) *The Intemperate Rainforest*. Minneapolis: University of Minnesota Press.
Burton, R., Kates, R. and White, G. (1978) *Environment as Hazard*. London: Oxford University Press.
Callon, M. (1986) 'Some elements in a sociology of translation', in J. Law (ed.), *Power, Action, Belief*. London: Routledge. pp. 19–34.
Castree, N. and Braun, B. (eds) (2001) *Remaking Reality*. London: Routledge.
Castree, N. (2003) 'Geographies of nature in the making', in K. Anderson et al. (eds), *Handbook of Cultural Geography*. London: Sage. pp. 1168–83.
Castree, N. and MacMillan, T. (2003) 'Dissolving dualisms', in N. Castree and B. Braun (eds), *Social Nature*. Oxford: Blackwell. pp. 208–24.
Coronil, F. (1997) *The Magical State*. Chicago: University of Chicago Press.
Dean, M. (1991) *The Constitution of Poverty*. London: Routledge.
Dean, M. (1999) *Governmentality*. London: Sage.
Demeritt, D. (2001) 'Being constructive about nature', in N. Castree and B. Braun (eds), *Social Nature*. Oxford: Blackwell. pp. 22–40.
Drayton, R. (1996) *Nature's Government*. New Haven, CT: Yale University Press.
Eagleton, T. (2000) *The Idea of Culture*. Oxford: Blackwell.
The Economist (1997) 'The age of the chicken', 18 March.
Ferry, L. (1995) *A New Ecological Order*. Cambridge, MA: MIT Press.
Foucault, Michel (2000) *Power*. New York: The New Press.
Forsyth, T. (2003) *Critical Political Ecology*. London: Routledge.
Geidion, S. (1948) *Mechanization Takes Command*. London: Oxford University Press.
Giddens, A. (1990) *The Consequences of Modernity*. London: Polity.
Glacken, G. (1967) *Traces on the Rhodian Shore*. Berkeley, CA: University of California Press.
Goodman, D. (2001) 'Ontology matters', *Sociologia Ruralis*, 41 (2): 182–201.
Greenblatt, S. (1991) *Marvelous Possessions*. Berkeley, CA: University of California Press.
Gregory, D. (2001) 'Postcolonialism and the production of nature', in N. Castree and B. Braun (eds), *Social Nature*. Oxford: Blackwell. pp. 84–111.
Grossman, L. (1984) *Peasants, Subsistence Ecology and Development in the Highlands of Papua New Guinea*. Princeton, NJ: Princeton University Press.
Haas, P. (1990) *Saving the Mediterranean*. New York: Columbia University Press.
Hajer, M. (1995) *The Politics of Environmental Discourse*. London: Clarendon.
Hall, S. (1992) 'The west and the rest', in S. Hall and B. Gieben (eds), *Formations of Modernity*. Cambridge: Polity. pp. 23–48.
Haraway, D. (1992) 'The promises of monsters', in L. Grossberg et al. (eds), *Cultural Studies*. London: Routledge. pp. 275–332.
Harre, R., Brockmeier, J. and Mühlhauser, P. (eds), (1999) *Greenspeak*. London: Sage.
Harvey, D. (1974) 'Population, resources and the ideology of science', *Economic Geography*, 50: 256–77.
Harvey, D. (1996) *Justice, Nature and the Geography of Difference*. Oxford: Blackwell.

Harvey, D. (2003) *The New Imperialism*. Oxford: Clarendon.
Henderson, G. (1999) *Fictions of Capital*. London: Oxford University Press.
Jassanoff, S. (1990) *The Fifth Branch*. Cambridge, MA: Harvard University Press.
Kapucinski, R. (1982) *Shah of Shahs*. New York: Harcourt.
Kautsky, K. (1899) *The Agrarian Question*. London: Merlin.
Koerner, L. (1999) *Linnaeus: Nature and Nation*. Cambridge, MA: Harvard University Press.
Latour, B. (1987) *Science in Action*. Cambridge, MA: Harvard University Press.
Latour, B. (1991) *We Have Never Been Modern*. Cambridge: Harvard University Press.
Leach, M. and Mearns, R. (1997) 'Challenging the received wisdom in Africa', in *The Lie of the Land*. London: Currey.
Lewontin, R. (1989) *The Triple Helix*. New York: Harper.
Lewontin, R. (1982) 'Organism and environment', in H. Plotkin (eds), *Learning, Development and Culture*. Chichester: Aldgate.
Lewontin, R. (1993) *Biology as Ideology*. New York: Dimension.
Lewontin, R. (2001) *The Triple Helix*. Cambridge, MA: Harvard University Press.
Luhmann, N. (1991) *Ecological Communication*. Stanford, CA: Stanford University Press.
Mann, S. and Dickinson, J. (1978) 'Obstacles to the development of capitalist agriculture', *Journal of Peasant Studies*, 5: 466–81.
McNally, D. (1995) *Against the Market*. London: Verso.
Merchant, C. (1980) *The Death of Nature*. New York: Harper and Row.
Mitchell, T. (2000) 'The stage of modernity' in T. Mitchell (ed.), *Questions of Modernity*. Minneapolis: University of Minnesota Press. pp. 1–34.
Mitchell, T. (2002) *The Rule of Experts*. Berkeley, CA: University of California Press.
Moore, D. (2005) *Suffering for Territory*. Durham. Duke University Press.
Neumann, R. (1999) *Imposing Wilderness*. Berkeley, CA: University of California Press.
Nietschmann, B. (1972) *Between Land and Water*. New York: Academic Press.
Ollman, B. (1972) *Alienation*. Cambridge: Cambridge University Press.
Polanyi, K. (1967) *The Great Transformation*. Boston, MA: Beacon.
Pollan, M. (2001) *The Botany of Desire*. New York: Knopf.
Povey, M. (1998) *A History of the Modern Fact*. Chicago: University of Chicago Press.
Raffles, H. (2002) *In Amazonia*. Princeton, NJ: Princeton University Press.
Rappaport, R. (1968) *Pigs for the Ancestors*. New Haven, CT: Yale University Press.
Reichel-Dolmatoff, G. (1972) *Amazonian Cosmos*. Chicago: Aldine.
Rose, N. (1999) *Powers of Freedom*. Cambridge: Cambridge University Press.
Rothfels, N. (2002) *Savages and Beasts*. Baltimore, MD: Johns Hopkins University Press.
Sayer, A. (1997) 'The dialectic of culture and economy', in R. Lee and J. Wills (eds), *Geographies of Economies*. London: Arnold. pp. 16–26.
Schama, S. (1995) *Landscape and Memory*. New York: Knopf.
Schurmann, R. (ed.) (2003) *Remaking the World*. Berkeley, CA: University of California Press.
Sebald, W. (2003) *After Nature*. New York: Modern Library.
Serres, M. (1990) *The Natural Contract*. Cambridge, MA: Harvard University Press.

Shiva, V. (1988) *Staying Alive*. London: Zed Books.
Slater, C. (2002) *Entangled Edens*. Berkeley, CA: University of California Press.
Smith, N. (1990) *Uneven Development*. Oxford: Blackwell.
Soper, K. (1995) *What Is Nature?* Oxford: Blackwell.
Stepan, N. (1991) 'Tropical nature as a way of writing', in A. Lafluente et al. (eds), *Mundializacion de la Ciencia y Cultura National*. Madrid: Doce Calles. pp. 495–504.
Stepan, N. (2001) *Picturing Tropical Nature*. Ithaca, NY: Cornell University Press.
Swift, J. (1997) 'Narratives, winners and losers', in M. Leach and R. Mearns (eds), *The Lie of the Land*. London: Currey. pp. 73–90.
Swyngedouw, E. (1999) 'Modernity and hybridity', *Annals of the Association of American Geographers*, 89 (3): 443–65.
Taussig, M. (1980) *The Devil and Commodity Fetishism in Latin America*. Chapel Hill, NC: University of North Carolina Press.
Wadell, E. (1975) 'Frost over Niugini', *Human Ecology*, 3: 249–73.
Whatmore, S. (2002) *Hybrid Geographies*. London: Sage.
Whiteside, K. (2002) *Divided Natures*. Cambridge, MA: MIT Press.
Wilden, A. (1972) *System and Structure*. London: Tavistock.
Williams, R. (1976) *Keywords*. Oxford: Oxford University Press.
Williams, R. (1993) *The Sociology of Culture*, New York: Schocken.
Wilson, A. (1991) *The Culture of Nature*, Toronto: Between the Lines Press.
Wolf, E. (1977) *Europe and the People Without History*. Berkeley, CA: University of California Press.
Young, R. (1995) *Colonial Desire*. London: Routledge.

9 Local: Global

Kevin R. Cox

Context

The attempt to grasp difference through the use of a local/global contrast might appear to be a 'given' for human geography. Scale has been a major preoccupation of the field and the local/global is one way of approaching that issue. But in fact, and until recently, resort to this particular binary was notable for its relative absence. One has to go back to 1919 and Mackinder's *Democratic Ideals and Reality* to find something similar to what has recently become such common currency. This suggests, among other things, that it needs to be approached in terms of its socio-historical constitution. In other words, and as part of a critical approach, it needs to be contextualized. In this regard it would be hard to resist the conclusion that its recent efflorescence owes much to the widespread interest in 'globalization' and its discontents. This is not to overlook, however, that the separation it implies is a chronic feature of the sort of consciousness conditional upon capitalist development; something it shares with other dualisms.

This is in addition to the other problems that Sayer (1991) has identified with binaries. One of these is the way in which they tend to get aligned with still others. In the local/global case, and again reflecting its recent more general usage, these have included state/capital, national/international, immobile/mobile, labour/capital, concrete/abstract and even agency/structure. As we will see, these alignments are oversimplifications of an almost heroic magnitude. In consequence, one of the ways in which the globalization debate has developed has been through a softening of contrasts. The other point that Sayer makes is that the relata defining a dualism are not necessarily just in opposition but may also be mutually presupposing; they are necessary conditions for each other. Again, making claims of this sort has been part of the recent history of the local/global binary.

Having said all this, there is also a certain distinctiveness attached to the local/global which stems from the different ways in which we typically talk about scale. Contemporary references to globalization almost invariably imply the literal 'global'. The 'local' is a little more troublesome, since it could apply to either 'the national' or something at the sub-national scale. Already, by using the term 'national' some difficult issues are entering into the discussion. But the one I want to draw attention to here is the difference between what I will call absolutist approaches to defining scale on the one hand, and relativist ones on the other. There are conventional usages which define the local and the global as literally that. But there is also a more relativist definition in which the global simply refers to the wider field of events and conditions significant to what is going on at a smaller geographical scale – which could, in a more absolutist sense, be a city, a region, or even a neighbourhood. There are more concrete uses, therefore, which need to be referred to the geohistorical, as in the contemporary interest in globalization as something that is international. But a critical approach to those concrete uses needs to be rooted in a more abstract understanding which, in turn, is a product of less specific, more general conditions, that is, those defined by capitalism.

The chapter is divided into three major parts. In a first, and relatively short section, I want to review briefly the relation between the local/global and the history of geographic thought in the twentieth century; precisely what is it about that history that helps us understand why this particular binary has, until recently, been notable more for its absence? In the second section I take up the relation between the local/global and the ongoing debate about globalization. Arguably it is this debate that has been so important in projecting the local/global distinction onto academic as well as lay consciousness. The way in which globalization and its implications have been conceptualized, moreover, provides provocative materials for a disinterring of precisely what the local and global might be assumed to mean both in contrast to each other and in their mutual relationships. This discussion then provides the context for the final and major section of the chapter. Here I try to situate the local/global with respect to the idea of spatial imaginaries: ways of conceiving space, that is, which have political intent. These are a necessary concomitant of the tension-ridden path of capitalist development because that development has inherently geographical moments: it occurs not just through a socialization of production but through a socio-*spatialization* of production; and the inequalities it necessarily throws out include the geographical. These arguments are then illustrated through a contrast between the spatial imaginary of globalization which is so dominant today and the one that preceded it during capital's so-called 'golden years': an imaginary which conceived of the economic dilemma in terms less of state failure and more those of market failure. Space was seen as something to be planned and rationally organized rather than a force in itself to which people had to adapt.

The local/global in the history of geographic thought

A common way of organizing thinking about the variety of ways in which human geography has been approached is in terms of the so-called 'three traditions'; the chorological or regional, focusing on areal differentiation; the environmental, with its emphasis on people's relations with nature; and the spatial, in which relations over space are highlighted. The relative predominance of these different ways of framing geographies has clearly changed over the past one hundred years. In the opening years of the century people–environment relations provided the principal framework for organizing geographic knowledge and understandings. The supposed excesses of environmental determinism, the concern that geography lacked 'an object of study', then led to a revival of interest in, a reworking of the notion of, the region, though the environmental continued as an important sub-theme. It was against this combination, but more against the emphasis on regional particularity, that the spatial-quantitative revolution of the late 1950s and early 1960s set itself. This was largely on methodological grounds: the belief that arguments about the particularity of place notwithstanding, geography could be a generalizing science and the way to do this was through abstracting out the (comparable, measurable, therefore generalizable) spatial relations of geographically defined observations: cities, factories, farms, households and the like.

Throughout this period there was obviously an interest in geographical scale. But it tended to be defined as 'large' *vs.* 'small', or as macro-, meso- and micro-, rather than in terms of the relations between entities organized at different geographical scales, which is arguably the essence of the local/global. One might have thought that the spatial-quantitative revolution would have changed that. But, on the whole, and for a variety of reasons, it did not. Partly it was because 'local' was subsumed to 'place' and hence to the uniqueness which was a part of the geography that was being contested. Space/place was an important binary in the era of the spatial-quantitative revolution, though the attempt was always to see place as a residual which might in turn give clues as to how it could be assimilated to the spatial and therefore the generalizable. The other problem was the reliance on micro-economics for its social theory, something entailed by the role of statistics in the spatial-quantitative revolution and hence the notion of independent observations. Concepts of organization and organization over space were, as a result, weakly developed.[1] The world of human geography was a world of individuals, or quasi-individuals: people commuting and migrating, firms relocating, households choosing residential neighbourhoods.[2] In consequence, and with one exception, approaches to scale questions progressed little further than exercises in variance decomposition.[3] The exception was the recognition that the population density affected the geographic scale of movement and hence the scalar organization of the space economy. This was one of Berry's (1967) insights coming out of his work on central place theory, later to find a more concentrated theoretical expression in the work of Curry (1972) on gravity coefficients. But the way in which longer distance movements might articulate with the shorter was never on the agenda.

It was only subsequent to what one might call the 'socialization' of human geography, a reaction in turn to the spatial-quantitative revolution, and more particularly the realization that people, industrial plants, retail outlets etc. were linked together in identifiable organizational forms that had a geographical expression, that a space was created for the local/global, though even then it did not happen till the broader debates about globalization. In this regard the Marxist moment in human geography was a significant development, though it influenced many others who would reject that affiliation. Marxism was initially the dominant vehicle through which the spatial became the *socio*-spatial. It is altogether appropriate, therefore, that we can find the roots of the local/global distinction, though not yet articulated as such, in the work of two of the most fertile thinkers in that genre: Doreen Massey and David Harvey. Massey's contribution was in large part through her (1984) work on corporate organization and the associated spatial divisions of labour of firms. Through these ideas she emphasized the stretching of social relations over space and how something organized at a supra-local scale could exercise effects on the functioning of more local-level forms of organization, including (through its impact on gender relations) that of the household. Harvey's insight came through his attempt to spatialize Marx and his recognition that the use value/exchange value contradiction at the heart of capitalist development had its geographical counterpart in the form of the contradiction between space as a use value and space as an exchange value. This in turn set up struggles between cross-class alliances in particular places as they tried to mobilize the more global forces of capital mobility in defence of their place-specific investments. From these two stepping stones, and once 'globalization' became the flavour of the day, it was a very short jump to thinking in terms of an opposition between the local and the global.[4]

Before we talk about the globalization debate, however, and to qualify my argument that the local/global contrast has only recently sprung upon the geographic scene, I want to point out an exception that, as I hope to show later, will in turn serve to prove a rule. I refer here to the point I made in the first paragraph of this chapter about Halford Mackinder's (1919) *Democratic Ideals and Reality*. In the light of contemporary debates, his discussion there is of extraordinary interest. His book is a post-World War I tract setting out principles for national and international reconstruction, though it is better known, of course, for its enunciation (once more) of the heartland theory. For Mackinder, the problem that had led the world into war and which he believed would do so again if it remained unchecked, was the world market and the incentive structures it had created. Nations had specialized, and through trade and the subsequent struggle for markets, this had led them into conflict with one another. The need, therefore, and something that echoes down to the present day in the form of protectionism, was for what he called 'balanced development' on the part of states (1919: 234).

Within nations the imbalance brought about by the market was, he believed, resulting in the organization of people into classes, creating the prospect of organization across international boundaries and 'the idea of class warfare between the international proletariat and international capitalism' (p. 240). The answer once again was 'balanced development': 'the one thing essential is to displace class organization, with its battle cries and merely palliative remedies, by substituting an organic ideal, that of the balanced life of provinces, and under the provinces, of the lesser communities' (p. 242). But for the most part the local/global contrast was missing from the vocabulary of twentieth-century geography. The context of intellectual debate which changed all that should therefore be a necessary object of our concerns.

The local/global and the 'globalization' debate

A major vehicle for the congealing of the local/global distinction within human geography and providing a context for a critical understanding of it has been the debate that has emerged over the past 10 or 15 years around what has become known as 'globalization' and its various economic, cultural and environmental implications. Given its taken-for-granted nature it is easy to forget just how recent the interest in globalization and its implications for the so-called 'local' is ('so-called' because it has been used in argument over the past 15 years or so to refer to a number of different geographic scales, ranging from the localities of locality studies to the nation-state that has been the primary locus of the 'local' in more recent debates). It would be difficult to say precisely when the interest in the global and the local began to gel as a couplet. My guess is that it occurred some time in the early to mid-1980s. Before that there was not much evidence for it; rather issues that were later to be referred to as 'globalization' were interpreted in a relatively aspatial manner. In particular, there was quite a bit of writing on the future of the welfare state that had yet to make the connections that are now assumed as given. Since then, of course, reams have been written about it. It has insinuated itself into political debate, domestic and international. It became a key concept of the late twentieth century, always drawing meaning from its relations with events at smaller geographic scales, particularly the national. It continues as such beyond the millennium and shows no sign of slackening.

Nevertheless, the claims that have been made about it have clearly shown some degree of evolution. There have been various corrections, rebuttals and counter-arguments. This has been so both with respect to the way the relations between the two terms have been understood as well as their respective meanings. There was an initial, highly over-generalized phase that identified as key relations the supposed effect of globalization in its economic dimensions on states and on labour. Capital, it was claimed, had become hyper-mobile, roaming the world in search of the most profitable outlets.

This had produced a problem of territorial non-coincidence (Radice 1984) for states; their ability to regulate respective national economies on behalf of growth and full employment had been severely compromised by the expansion of foreign direct investment, the internationalization of financial flows, the expansion of trade and the rise of the Newly Industrializing Countries (NICs) of the Far East. At the same time, there were distributional implications. Lacking the international mobility of capital, labour – or at least labour in the North – found that its bargaining power had diminished and that in order to hang on to jobs or attract new ones in the form of inward investment, new and less favourable contracts had to be negotiated, benefits reduced and a contraction of the welfare state accepted.[5] Or so the argument went.

For it later became clear, to at least some, that this was not exactly what had been happening. More nuanced accounts then began to emerge. Some of these dwelt on the continuing power of the national, the ability of states to find roles in the changed context, to substitute microeconomic for macroeconomic interventions (Weiss 1997). This might be achieved, for example, by shifting from the 'low road' of cost competition to the 'high road' defined by developing new products and technologies which could provide, at least for a while, a quasi-monopoly and so underwrite the continuing prosperity of those regions or countries dependent on the firms in question, particularly if, as it began to emerge, those firms might not be as mobile as those that had generated images of hyper-mobility. Other work focused less on the question of the continuing ability of states to make a difference and more on that of localities, as in the British 'localities research' programme.

At its most ambitious, this particular thrust of critique reversed the causal arrow between the global and the local. Rather than the global making the local, the reverse applied. This was the gist of an early intervention by Storper (1987), though less in the context of the globalization debate and more to do with the future of Marxism in what he defined as a post-Enlightenment era. His point was that if we were to understand the development of capital as a worldwide force we had to look to what he called 'historically-, geographically-specific experiments' which then got 'totalized through adoption, diffusion and enforcement in real capitalist societies'. As he remarked, 'The local makes the global at certain critical points in space and time. I would argue that capital is forever altered by the rise of Japan, not simply revealed "in another instance"' (1987: 420). In short, globalization was, among other things, about the adoption of 'best practice' and 'best practice' always had its origins in the local.

A third alternative, of course, was to see the relation as a reciprocal one: the global and the local as mutually constitutive. According to this view the global was formed through the activities of many different localized agents; but then through both its limiting effects and through the possibilities it opened up for them, it conditioned their further development. There is a sense of this in Scott's work on agglomeration: the exploitation of increasing returns to scale in areas characterized by vertical disintegration of firms and a deepening division of labour forcing out competition elsewhere and so furthering the agglomeration process (for example, Scott 1996).

One interim conclusion that we might draw, therefore, is that the debate has been conducted in highly reified terms. The local and the global are conceived as separate objects which impact one upon another, much like the proverbial billiard balls of anti-positivist argument, though what impacts on what depends on the position taken. And as objects, of course, they are assigned attributes so that the global/local dualism has tended to get aligned with numerous others, along the following sorts of lines:

local/global
agency/structure
national/international
state/capital
immobile/mobile
labour/capital
cultural/economic
concrete/abstract
contingent/necessary

Sometimes this has been more implicit than otherwise. That the global is the sphere of the international, for example, has often been taken for granted. Moreover, and more importantly, these alignments have themselves been the object of a critique, and that critique has in turn been one of the ways in which the debate about the global has been moved forward, albeit without forsaking the highly reified terms in which it has occurred. Consider the following examples.

Labour/capital As I pointed out above, a common initial claim was to read into the global/local relation a shift in the employment relation. To capital was assigned a high level of mobility, typically international. This was referred, *inter alia*, to a de-skilling of labour processes that allowed the decanting of production functions to low wage areas of the world and to changes in transportation and communication; the former in the shape of innovations like the container allowed what was produced offshore to be shipped back to markets in the so-called Triad countries; while the latter facilitated the coordination of the far-flung constellations of the plants and offices of the multinationals. Labour, on the other hand, was seen as much more immobile by virtue of a range of considerations including family and the catch-all, 'culture'. Interestingly, the notion that, given the income supports of the welfare state, they could not possibly have benefited themselves by following the multinationals overseas seems to have gone unremarked.

What was remarked on, however, as the globalization debate moved beyond its initial, highly over-generalized stage, was the fact that capital could not be pigeonholed in this way; that while some parts of the labour process for some products were indeed mobile to the degree that they could be relocated to low wage areas of the world, other parts certainly could not be. American multinationals might indeed close plants in the US and shift production to a *maquiladora* in Mexico or subcontract to firms in the Far East but most of their assets and sales would in all likelihood remain in what Hirst and Thompson (1996) called their "home base". As Michael Mann put it (1997: 479): 'The ownership, assets and R&D of "multinational" corporations remain disproportionately in their "home" state, and they still lean on it for human capital (education), communication infrastructures and economic protectionism'.[6]

This should not have been a surprise. The de-skilling of labour processes has always been highly uneven so that multinationals would continue to rely on the Triad countries for, as Mann put it, 'human capital'. But even where they could be relocated to LDCs, quality control issues might be a deterrent owing to the lower frequency of deliveries of parts to more developed countries, even if transportation was cheaper in real terms. In other words, more frequent, smaller deliveries allowed a feedback of information to suppliers that was not possible where transport lines were more stretched out. In some cases trade barriers have interfered with the relocation of production, which helps explain why the American computer firms were able to obtain semiconductors more cheaply than their West European counterparts (Sayer 1986). And on top of that some firms are just too small to relocate; they lack the deep pockets necessary to underwrite what is essentially a risky venture.

Of course, to the degree that the jobs in the more developed countries were not replaced this could still alter labour market conditions there, just not as much as under the more catastrophic scenarios of a race to the bottom described in earlier versions of the globalization thesis.[7] What had been left out of the equation, however, and what still tends to be left out, was the dynamic character of the division of labour; how it was not a matter of redistributing a set number of concrete labour processes or parts of those labour processes, but of dividing labour that was itself changing in terms of concrete products and services. In other words, the decanting of some production to low wage zones facilitated the development of new branches of production in the more developed world. It helped increase discretionary spending there on the basis of which new industries could emerge at the same time as it released labour for them.

State/capital But if capital has been recognized as both immobile and mobile, thus weakening some of the supposed contrast between immobile labour and mobile capital the state has, for the most part, tended to remain stubbornly national in these accounts. Again, this is a distinction that should be softened. Even prior to the hype about globalization, in fact for quite some time, it was clear that state forms were changing and the national element was of decreasing significance. The obvious change here is the emergence of diverse supranational organizations such as the EU, NAFTA, the WTO and its predecessor GATT, the Bretton Woods agreement, and so on and so forth. In other words, and as Shaw has argued, the state has also been a globalizing institution.

Indeed Shaw has gone so far as to argue for the emergence of a distinct 'Western' state which he defines as 'a massive, institutionally complex and messy agglomeration of state power centered on North America, Western Europe, Japan and Australasia but whose writ ... has had in many senses genuinely global reach' (1997: 503). This, moreover, has been a precondition for globalization in the more recent, economic, sense since it provided the political-military framework for it. Some might argue that this reorganization does in fact imply a loss of capacity on the part of the nation-state, but whether this is a good or a bad thing – and the globalization literature has tended to assume that it *is* a bad thing – is something that needs to be addressed in sober fashion; and in answering that one has to ask, for *whom* might it be good or bad, a question we will return to later in the chapter. Having made these points, however, it is evident that the critical dissection of the local/global binary and its various associated dualisms can be taken a good deal further than through a discussion of issues of relative mobility or the state/capital contrast. I conclude this section, therefore, with two further examples.

Agency/structure In the first place there has been some tendency to align the local with agency and the global with structure. This is evident in the work of Linda Weiss, for example, and also in the paper by Michael Storper which was referred to above. The problem is, however, that the global is always manifest in the form of agents, albeit structured agents, as in the cases of (for example) particular multinationals or international organizations or firms that have established geographically extensive distribution networks. Likewise, to equate the local with agency would be to ignore the way in which local agents are structured in their activities, and not just by 'the global', even when it is manifest in the more concrete forms of particular connections with concrete places and organizations, firms, credit rating agencies and the like. Rather more global forces are a threat because of the way in which firms are often embedded in local social relations: labour markets, supplier networks. Yet that structure of social relations can also empower, perhaps in terms of a quasi-monopoly in the production of a particular commodity, and provide points of leverage through which the local in its diverse manifestations can make a difference. So local agency, while always and necessarily structured, since agency always has its conditions, is not necessarily structured exclusively by 'the global'. In short, a more defensible position would be one in which the agency/structure contrast is recognized as orthogonal to the local/global one.

National/international This dualism is a very curious one. One way of looking at the globalization of the economy, for example, is in terms of the scale division of labour: the geographic scale at which labour is divided. In the case of the United States this has been, for the most part, national: the US happens to be a relatively closed economy. While the United Kingdom imports timber from Norway, fresh off-season fruits and vegetables from Spain and Italy, and sends its tourists to the Mediterranean, for an American state like Illinois these functions are fulfilled from within the national boundaries. For West European countries, therefore, globalization in the sense of the elongation of commodity flows has meant the creation of a single European market with accordant adjustments in the scale at which labour is divided. Significantly, the trade of the member countries of the European Union as a percentage of EU product is now lower than the equivalent figure for the US (Kleinknecht and ter Wengel 1998). What is interesting, however, is that the tensions that are now emerging in the EU between the 'national' and the 'international' found their equivalent many years ago in the US. In other words, in the US the creation of a national economy, the widening of the scale division of labour to correspond to the national, generated the same sorts of tensions over uneven geographic development and the competition for investment that are now characterizing the newly open national economies of Western Europe. This suggests that our understanding of the local/global could usefully be seen in abstraction from the national/international, and more particularly that the national/international bears no necessary relationship to the local/global: in the US case for the individual states the global has for a long time been the national; in the case of the member countries of the EU the global is the international in the form of the EU.

From concrete to abstract and back again: towards a critical understanding of the local/global

The local and the global in the abstract

The first point to make here is that in deconstructing the local/global binary we will find it useful if we extricate it from its more recent and more specific use; that is, the use which I have discussed above and which has received so much media and academic attention. The way it is called upon in that context should give us pause anyway, since, more often than not, and as I have pointed out, the substitution of a national/international distinction would suffice as a framework for the discussion. Rather, the local/global distinction should be seen as thoroughly generic and not limited to the particular geographical scales highlighted by the globalization debate. In other words, there is leverage to be gained by viewing the local/global difference in terms of a relativist as opposed to an absolutist conception of geographic scale. According to this the local can assume all manner of (absolute) scale expressions, from the neighbourhood, through the city, region and nation, even the continent, so long as the global which it implies and to which it relates through substantive connections is indeed at a larger geographical scale.

For sure the terms 'local' and 'global' may not be used, but the way in which the particular terms that are used are deployed is thoroughly homologous, as for example when neighbourhood change is situated with respect to a wider metropolitan housing market. And in some cases, while the term 'local' may indeed be used, as in locality studies, the wider global within which localities may or may not 'make a difference' may not be, and in fact in that instance was not, specified; so in absolute terms it could have been national or international or both at the same time. The old notion of national integration smacks profoundly of a local/global relation, and indeed generated, and continues to generate in large parts of the world, including the North, the same sorts of tensions identified by the globalization debate; but again the global being specified in those instances would not be identified as such.

At least, not always. And this suggests another advantage of viewing the local and the global in both more abstract and more relativistic terms. If there is more than one local and more than one global, if what is global depends on the circumstances, then on any particular occasion more than one may be called on to do service for whatever interests are at stake. Something like the spatial imaginary conjured up by the Northern League in Italy brings this strikingly home since their projected state of Padania is situated with respect to several globals: not just Italy and those parts of the Mediterranean funnelling immigrants into Italy, but also the EU, for getting into Euro-land was as much part of the agenda as escaping from the Roman state. More generally, the mobilization of differences of interest between the different globals has always been a weapon in the arsenal of local interests resisting the state, ranging from struggles against apartheid to the more recent bypassing of nation-states in favour of links with the EU. The way in which localities mobilize a 'less global' global in order to achieve advantage with respect to a 'more global' global, as in the pressure brought to bear on national governments for protection of local industries, is also common enough.

The local/global as a spatial imaginary

So the local/global relation, even though it may not be identified as such, is certainly a common feature of the way space, along with its politics, is imagined, and at diverse geographical scales. It is, in other words, one of the ways in which people try to make sense of the dilemmas that inevitably arise within a capitalist society as its geographic contours change, including its scale divisions of labour and consumption. But at the same time we need to recognize that it is an abstraction from more complex socio-spatial relations; a construction, therefore, which simplifies, silences and can be turned to useful ideological effect. Spatial imaginaries, like the local/global, are always constructed from the standpoint of particular social positions, if not always in a very self-conscious manner.

To start with, consider the characteristic ways in which capitalist space economies develop, the contradictions that emerge, the practical and discursive ways in which those contradictions are confronted with a view to their suspension. Central to capitalist logic is the pursuit of value as a thing in itself, regardless of its preconditions in actual material relations: those of the labour process and of the reproduction of labour power, for example. This is rooted in the seeming separation of exchange from use value that comes about with the development of the money commodity and the all-round dependence on exchange that occurs subsequent to the separation of immediate producers from the means of production. But this separation of exchange value from use value is manifest in other semblances latent with crisis. These include the separation of the individual from society and of labour from capital.

Geographically, the unity of opposites that is the commodity is expressed in the form of a space as a use value alongside space as an exchange value. The development of capitalist space is part and parcel of the separation of use value from exchange value. An earlier concrete space tied down to specific labour processes and the spatial arrangements conditioning them co-exists with, is subordinated to, is re-worked by, an abstract space defined by the flow of value. This is the point of the emphasis that Harvey gives to the relation between fixity and mobility: in particular, to the contradiction between them and how that contradiction, as a particular expression of the use value/exchange value contradiction generates a distinctive geography, a particular unity of society and space.[8] As Harvey describes at length, if value is to be appropriated it has first to be sunk in fixed facilities, physical infrastructures and networks of relations with others have to be constructed: supplier networks, an appropriate labour supply. But once appropriated it becomes mobile once more and is free to seek out more profitable investments elsewhere. Furthermore, given the class contradictions of capital, the way firms seek to suspend that contradiction through the development of new technologies, new products and forms of organization, it is likely that it will. In substantive terms the mobility of value in the money form threatens value locked up in various fixed forms: factories, workers' housing, utility networks. It also imposes opportunity costs in the form of abandoned social infrastructures that were built up slowly and painstakingly.

It is not difficult to see that one of the ideological reflections of this contradiction might well be a sense of a local oppressed by more global forces, and this of course has found expression in the globalization debate. Equally, the opposition can be one in which the global is seen as something that can be turned to local advantage (attracting inward investment, mobilizing innovation, selling the area as an exchange value) and this too is an interpretation that has found common currency. But all these views, the local producing the global, the global forming the local, the global as a condition, depend on reified views of space. These in turn depend on that separation of use from exchange value which, as Harvey pointed out, receives a distinctly geographic expression.

This is not all that is going on. For sure, capital develops through a socialization of production and a major, if contingent, expression of this is the creation of wider spheres of exchange; it is therefore a deepening of the interdependence between producers in particular places, and so a deepening of the contradiction between space as exchange value and space as use value. But along with that socialization of production goes an increasing inequality: between capitalist and worker, between different positions in the 'collective labourer'. This too acquires a geographic form in a profound differentiation of localities: in particular, of different roles in the geographic division of labour, both social and technical – rustbelts *vs* sunbelts, branch plant towns *vs* corporate headquarter and R&D towns; white collar places *vs* blue collar places, more developed as opposed to less developed countries. This means that another source of tension can be added to that between the local and the global generating in its turn claims of territorial exploitation, of colonialism, and of the need to 'level the playing field'.

This says nothing of the geographical scale at which these processes play out: the scale of particular locals/globals or of spatial divisions of labour; that is, the scale at which labour is divided. This, it needs to be emphasized, and as Radice (2001) has recently underlined, is always a contingent matter. The degree to which the ongoing socialization of production is expressed through expansion beyond those spaces in which value has circulated hitherto, or alternatively through a reorganization of existing spaces, an intensification of their uses, perhaps, is not something that can easily be anticipated. There is a sense in the writings of many that globalization is a secular process knitting producers together over ever-increasing distances. Yet as a number of people have pointed out, there is also a sense in which national economies were *more* globalized at the turn of the century than they are now. Mackinder's concerns did indeed reflect a period in which there was intense competition for international markets. But subsequently, including during the 1950s and 1960s – capital's so-called 'golden age' – national economies became considerably more closed and it is only since the early 1970s that there has been a shift back towards openness. Even then, the levels of openness characteristic of the opening years of the last century still have yet to be achieved.

There are also risks of over-spatializing the analysis – a common tendency which should be resisted. Radice (2001) reminds us of this in his recognition that capitalist economies can expand through intensification within an existing space rather than through processes of spatial expansion. Spatial strategies are contingent. What is fundamental is the class relation. Capitalist development is mediated by processes of class struggle, the reproduction of class relations that work to capital's advantage and which *must* work to its advantage if accumulation is to continue. This is not to ignore the relation between the individual capitals. It is this which enforces the logic of exploitation and accumulation (Weeks 1981). At the same time it mediates the distribution of the surplus thus appropriated. So the 'global' that capitals complain about consists of other firms which threaten their profitability. Moreover, to the extent that these differences acquire an areally differentiated form, charges of colonial relations, of territorial justice, can acquire a semblance of plausibility. But ultimately, what counts is the class relation: without the domination of workers by capital there is no surplus to fight over. So the material relations that give spatial imaginaries like the local/global, and those of territorial justice, their plausibility are not given by spatial dilemmas but by class dilemmas, albeit in the context of processes that have a strong spatial moment.

They are, furthermore, clearly imaginaries that obfuscate. At best they encourage the working class to believe that salvation lies in some new spatial fix, or the restoration of an old one, but in any case one that typically promises only to relocate the worst excesses of the class struggle. At worst, they paper over the cracks with notions of unity between capital and labour – Harvey's regional class alliances or, more ideologically, Mackinder's organic local communities: a unity that can vary in its choice of adversary from imperial forces to multinational corporations and Harold Wilson's 'gnomes of Zurich', and which through gambits like the white man's burden can even incorporate subject peoples. These are all possibilities latent within the capitalist form of development. But they remain just that: possibilities. We know that there will be contradictions and we know that they will assume a spatial form. What we cannot anticipate are the precise contours of the spatial imaginaries that will be called on to dispel the tensions and allow capital accumulation to continue. Much, therefore, depends on geohistorical conditions, how those contexts are interpreted, and how they are filtered and redefined as a result of the pressures and possibilities inherent in those contexts. It is to two examples of this that we now turn.

Applications

The globalization debate I would argue that the present interest in the local/global among academics owes a great deal to the larger intellectual context provided by the globalization problematic. This was a set of interpretations of the changing character of the global economy and its implications for more local geographical units, often national in character, but sometimes subnational, a number of which we reviewed earlier in this chapter. These were promulgated initially by politicians, media people and, in academia, the business school element. And importantly, they resonated in ruling class quarters. This was because of the way in which they provided a seemingly technical, apolitical understanding both of the crisis in which that ruling class found itself during the mid to late 1970s, and of the strategies that were being undertaken in order to resolve it. In short, the globalization problematic provided an intellectual response, which could be marshalled for more general consumption, to the contradictions that were being encountered in the further development of the capitalist economy and which would help clear away a path for their suspension.

The essential, if not immediate, context is provided by what Robert Brenner (1998) has called 'the long downturn'. This brought the so-called 'golden age of capitalism' of the 1950s and 1960s to a halt and was characterized by slowdowns in the growth of profitability, productivity, investment and employment. It resulted in large-scale, localized unemployment as a result of plant closures and also a more general increase in national unemployment rates, which for some countries have yet to find levels characteristic of the 'golden age'. The general policy response to the long downturn has been an attempt to reimpose the law of value through the implementation of a neo-liberal agenda. Globalization in the senses of establishing branch plants in the export processing zones of less developed countries, the shift to currency convertibility, the drive for export markets, has certainly been part of this; but, and I emphasize, only part. If we want to talk about geography there have been restructurings *within* countries, and ones not necessarily geared to competition in international markets. But there have also been, as is far too well known to dwell on at length, privatizations, the marketization of state functions, deregulation, an onslaught on labour through a rewriting of labour law and enhanced militancy in challenging organizing drives, and the move away from corporatist arrangements between state, labour and business. Nor should we neglect to mention the implementation of free trade areas.

However, if globalization has been only a part of this, it has been a crucial part. This is because of the essential role it has played in imposing the agenda discursively. Quite how the 'globalization discourse' emerged when it did would repay serious study. For it was not until sometime in the 1980s – long after the onset of the long downturn and the implementation of neo-liberal policies – that it crystallized. There was an earlier literature dwelling on the fiscal crisis of the state, the need to segregate production from distribution and the problem of governability that makes no mention of globalization;[9] and then, quite suddenly, there it is. What seems to have changed was the recognition of the difference it could make in defusing the class tensions that neo-liberalism inevitably generated. Attempts to drive down wages, which have certainly been part of the neo-liberal medicine, are inevitably politically fraught unless they can be referred to some anonymous force like 'international competition'. So too is it the case with creating mass unemployment in order to reassert the disciplines of the labour market. In short, arguments about a globalization juggernaut have provided a politically convenient cover for re-imposing the disciplines of the market, and in the highly diverse forms those strategies have assumed.[10] As Frances Fox Piven has written:

> The key fact of our historical moment is said to be the globalization of national economies which, together with 'post-Fordist' domestic restructuring, has had shattering consequences for the economic well-being of the working class, and

> especially for the power of the working class. I don't think this explanation is entirely wrong but it is deployed so sweepingly as to be misleading. And right or wrong, the explanation itself has become a political force, helping to create the institutional realities it purportedly merely describes ... Put another way, capital is pyramiding the leverage gained by expanded exit opportunities, or perhaps the leverage gained merely by the spectre of expanded exit opportunities in a series of vigorous political campaigns. (1995: 108–10)

The message internationally has been broadly similar. Mrs Thatcher's TINA applies to the modernizing elites of the less developed countries as much as it does to the working classes of the North. Without currency convertibility, without a drastic lowering of tariff barriers, a running down of the social safety net in the interests of budgetary balance and the protection of asset values, they risk marginalization and the loss of all possibility of growth in the future. So to the extent that resolving the contradictions of capitalism does indeed require the opening up of new export markets, and new opportunities for investment and takeover, then Third World elites have to toe the line, run their flags up the neo-liberal pole at the same time as they run it down on alternative ones like that of the Asian model or other such threatening deviations.

But along with rhetorics of objective necessity there have been ones of possibility, of empowerment. According to these, the globalization of economies, their opening up to international trade and investment, the reorganization of public finance in a business-friendly manner, define a configuration of forces which holds out possibilities of renewal/recapitalization/revival of the prosperity of the golden years, and a movement up in the worldwide geographic division of labour. The necessary policy initiatives may result in some short-term pain but eventually the benefits will trickle down, with the result that all boats will be lifted by the incoming tide.

The subsequent marginalization of alternative, class-based discourses, however, could not have occurred without a sympathetic conjuncture of class forces. In order that a particular discourse be acceptable and be capable of countering alternative ones, it has to respect the particular balance of political forces. In this case, it was not just a matter of a working class weakened by unemployment confronting a capitalist class determined that the costs of overaccumulation should be visited on its working class rather than absorbed, though this was important. There was also the fact of major sectoral shifts which resulted in the severe contraction of the most unionized sectors of the economy (in particular, coal mining, iron and steel, public transport, ship building) and a concomitant failure to replace the losses through organizing workers in the new, growth sectors of the economy. Complicating the picture even further has been the growth of female employment and the various problems posed to their unionization. These have included, first, the ideology of pin money,[11] and second, the historic neglect, even hostility, of the union movement to women in the wage labour force and to their elevation to positions of authority in that same movement. At the same time sectoral and geographic change in the economy has created new support bases for the forces of neo-liberalism. Under Mrs Thatcher the Conservative party strengthened its grip on Southeastern England as a result of the boom in financial services and to a lesser degree high-tech. In the US the Republican Party under Reagan cemented its domination of the Mountain West and the South in part as a result of the relocation of industry from the Midwest and the Northeast, relocations or shifts in investment that were a response, at least in part, to the threats to corporate profitability defining the long downturn.

In turn these geographic differentiations have created the conditions for alternative forms of spatial imaginary, ones in which there is room for the defence of working class privilege but at the expense of co-optation by local class alliances and at the expense of labour elsewhere. In the US an initial response to the emergence of the uneven development of Coldbelt and Sunbelt respectively were calls from local capitalist interests in the former for sacrifice on behalf of the local working class. Worker givebacks, changes in labour law, in state taxation, would lay the foundations for a recapitalization of localities. To some degree this worked. But given the fact that the Coldbelt is – or *was* – the stronghold of the American labour movement, there was also resistance. This in turn served to push local growth coalitions in the direction of an alternative strategy: one which argued not for working class concessions but for a levelling of the playing field and a redirection of federal fiscal flows.[12] In Britain the deindustrialization of the North and the emergence of the North/South divide led to similar protectionist responses so that, at least for a time, the social base of the Labour party assumed a strongly regional form (Savage 1987).

In short, capital has been able to take advantage of a moment of relative weakness on the part of the labour movement in the more developed countries. At the same time the changes set in motion by capital's mitigating strategies have created new social bases for the parties of the right and have served to divide labour into warring factions, albeit in alliance with respective local capitalist classes. This is in sharp contrast to the earlier period of globalization around the turn of the century and with respect to which Mackinder positioned his arguments. In the immediate pre-World War I period the labour movement was in the ascendant, achieved something of a climax with the Russian Revolution, and was only broken during the 1920s and 1930s by the forces of reaction and their ideologies of nation and church (Mann 1995). It was that militancy and the prospect of worldwide revolution that concerned Mackinder, so that rather than seeing globalization as an opportunity for capital, he saw it as an opportunity for labour and one that had to be countered through a retreat to what he called more balanced forms of national development and by the creation of organic local communities.[13]

The golden age Think back now to capital's golden age. The way in which the dynamics of capitalism and its space economy were interpreted were different, and to a quite substantial degree. The emphases then were not on letting the market rip, and on trickle down, but on state intervention. Market failure – rather than state failure – was a more common interpretation of economic dilemmas, and, as far as the space economy was concerned, urban and regional planning was the solution. Market forces, it was believed, resulted in over-urbanization, and in geographically uneven development. In order to counter these effects the state had to step in, creating new towns and redistributing employment away from the boom areas to those with relatively high rates of unemployment. There was some variation between Western Europe and the United States. In the US the major issue of uneven development was that of central city-suburban fiscal disparities and this lent impetus to programmes of urban renewal. More generally, in both North America and Western Europe growth was seen as far less problematic than it is today and so less a concern of public policy, including spatial policy, than equity, efficiency and macro-economic stability. Rates of economic growth were higher and the assumption was that this was a result of state intervention. This, moreover, lent further credence to the belief in the power of the state.

This was to change in the 1970s, as Harvey (1989) has noted. Urban managerialism gave way to urban entrepreneurialism. Issues of equity within metropolitan areas ceded space to ones of competing for inward investment. The top-down policies of central government yielded ground to local initiatives which brought the localities more into competition one with another. But for a while a different spatial imaginary had held sway: one which promoted the rational organization of space through state intervention, particularly *central* state intervention, and with a view to reducing uneven development and the inefficiencies that the market introduced into the space economy.

The balance of class forces underpinning this regulatory programme was accordingly a different one. The evident failures of the market in the 1930s weakened confidence in business solutions and enhanced support for the interventionist programmes of the left. The fact that these seemed to work, at least until the 1970s, and/or did not appear to interfere with accumulation, likewise weakened capital's resolve to oppose them. In light of subsequent events, however, it is reasonable to raise the question of the role played in all this by the, then, relative closure of national economies. According to the globalization problematic it was increased levels of international trade, of foreign investment and the achievement of currency convertibility that undermined the political power of the Western working class. Conversely, therefore, could it be said that the relative closure of national economies enhanced that power during capital's 'golden years'? This at least is the implication of a dominant form of globalization logic.

There are some problems with this line of argument. For sure, the form that globalization assumed, and in contrast to that in the early years of this century,[14] did indeed alter the employment relation through plant closures and unemployment. But as we have seen there were other strategies resorted to during the long downturn that had no necessary relationship to the increasing openness of national economies. These included technical change, shifts in the composition of employment across different sectors, the shift of employment to less unionized areas and fractions of the population, along with state policies of cutting back expenditures, the deregulation of transport and the privatizing of functions.

We can argue that there are shifts in the balance of class forces, therefore, but explaining why is more problematic. It is by virtue of the nature of their relation that capital and labour are locked in a conflictual embrace. It is the constant upward pressure on wages, exerted in many different forms, including simply movement between jobs, that forces capitals to constantly seek out more efficient ways of doing things and so to accumulate. As they release labour, so new sectors of production can come into being. And as successive revolutions in productivity raise real wages, so the purchasing power necessary to sustain the demand for the new products is released. Class struggle and accumulation are two sides of the same coin. But there are clearly periods in which this relationship works to labour's advantage: periods, that is, in which capital is willing to make concessions, either in terms of individual wages and conditions of work or in terms of the social wage. This is because they pose less of a threat to profitability. Markets are expanding, revolutions in productivity can hold real costs down, investment is booming, the subsequent fixed costs have to be recouped and that can only happen if plant is not idled by labour disputes, and so on. But just as clearly there are periods in which markets are not expanding, when perhaps productivity lags so that wage increases *do* threaten profitability.

Robert Brenner (1998) drew attention to the long downturn but has been roundly criticized for his explanation of it. For him it is a crisis brought on by overcapacity which in turn was a result of the heightened competition of capitals one with another (which has led some to attribute to him the same neo-Smithian Marxism he attacked in a seminal paper in *New Left Review*). For those who, unlike Brenner, have yet to abandon value theory, however, he has it the wrong way round. Rather it is overcapacity that has resulted in the intensity of competition experienced in the 1970s. That overcapacity in turn is related to the increased mass of value seeking profitable outlets relative to the availability of those outlets: an availability linked in turn to the tendency for the rate of profit to fall as labour is expelled from the labour process in the course of technical innovation. The alternative is the black box of long waves: the alternation of long, 25- to 30-year periods of steady growth with equally long periods of relative stagnation, like the succession of capital's 'golden years' and 'the long downturn'. But at any rate, understanding of shifts in the balance of power between capital and labour needs to look less to the contingencies of exchange relations and more to what is happening at the point of production.[15]

Concluding remarks

One cannot but be struck by the recency with which the local/global *as such* has entered the lexicon not just of geographers but of others, both academic and lay. It seems such an obvious way of making sense of experience. And to be sure, the concept of an inside and an outside in their mutual relations *is* a common one. One thinks, for example, of the notions of provincialism and cosmopolitanism as contrasting possibilities within that relationship. But as a means of grasping difference within the space economy or even cultural geographies, though my emphasis has been on the former, it is remarkably recent. In fact it now seems difficult to think of the local and the global except in terms of those associations: associations which first became apparent in the course of the ongoing debate about globalization, though as I point out above, there are important reasons why we should try to sever that particular link and see it more as a contingent one.

As a generic conception the local/global is rooted in experience of the capitalist space economy and the tensions attendant upon its development: in particular, the tensions between space as use value and exchange value respectively, though these tensions are expressions of the more fundamental one rooted in the commodity form and which, in turn, spawns those class struggles through which capitalist development occurs in its various aspects, including the geographic. But the degree to which it is drawn on as a way of making sense of events is a contingent matter. 'Sense' cannot be divorced from the ideological and therefore from the class tensions unleashed by capitalist development, the attempt to subordinate, to defuse and to co-opt; though that is not to say that capital has it all its own way and that particular spatial imaginaries, including the local/global, might not be the outcome of a class compromise and part of a broader strategy of shifting those costs of devaluation and unemployment onto other shoulders elsewhere.

The local/global has acquired such resonance in contemporary human geography that it is all too easy to think of it as a logical outcome of the development of a self-referential discipline. But like all binaries it is by no means pure intellectual artifice. Rather, it is closely bound up with the course of real world events, acquiring both meaning and significance through the concrete trajectories assumed by capitalist development and the struggles that are condition and outcome of those trajectories. Furthermore, it is through that mode of understanding that the analytical difficulties that it entails – ones of separation, reification, and overgeneralization in particular – are highlighted: difficulties that apply to all binaries.

FURTHER READING

The history of the local/global distinction within human geography remains to be written. Chapter 7 of Halford Mackinder's *Democratic Ideals and Reality* (London: Constable 1991) nevertheless remains a stimulating read and an important point of reference in constructing that history. Some landmarks in the recent upsurge of interest in the local/global include the British locality studies project. One of its outcomes was a collection of essays edited by Phil Cooke: Philip Cooke (1989) *Localities*. London: Unwin Hyman. Numerous other collections such as Johnston, Taylor and Watts's (2002) *Geographies of Global Change* (Oxford: Basil Blackwell), bear this imprint as well. Note also that the local/global distinction has become closely intertwined with interest in questions of scale. For a paper that exemplifies this relationship see: Kevin R Cox (1998) 'Spaces of Dependence, Spaces of Engagement, and the Politics of Scale', *Political Geography* 17, 1–24.

On the issue of binaries and their problems, a short paper by Andrew Sayer (1990) is excellent: 'Dualistic Thinking and Rhetoric in Geography'. *Area* 22, 301–303.

As I pointed out, recent interest in so-called 'globalization' has been immensely important for thrusting the local/global distinction into the consciousness of both geographers and a wider public. A very good critical introduction is provided by Wayne Ellwood's (2001) *The No-Nonsense Guide to Globalization*. London: Verso. Chapter 10 of my own (2002) *Political Geography: Territory, State and Society* (Oxford: Blackwell) is equally accessible. A more advanced discussion, as sceptical though less radical, is the book *Globalization in Question* (1999) by Paul Hirst and Graeme Thompson (Cambridge: Polity). A still more advanced and radical discussion can be found in Kevin R Cox (2004) 'Globalization, the Class Relation and Democracy', *GeoJournal* 60: 1, 31–41.

Paul Hirst and Graeme Thompson (1999) *Globalization in Question*. Cambridge: Polity Press.

Ronald J Johnston, Peter J Taylor and Michael Watts (eds) (2002) *Geographies of GlobalChange*. Oxford: Blackwell.

Sallie Marston (2000) 'The Social Construction of Scale', *Progress in Human Geography* 24: 2, 219–42.

Notes

1 With the exception, that is, of 'spatial organization' as 'spatial predictability or order'. It was, in other words, the notion of organization as something intentionally produced that was so poorly developed.
2 There are obviously some important exceptions to this, the most notable of which is the work of Hagerstrand and for whom the social in the senses of social networks and purposive, social organization over space were always central elements.
3 As in Haggett's *Locational Analysis in Human Geography* (1965) and Moellering and Tobler (1972).
4 One should also add here the recognition that markets themselves required purposive organization if they were to function.

5 There are many examples of this type of argument. Examples include; Burawoy (1985: 148–52), Gill and Law (1989) and Streeck (1992).
6 See also Epstein (1996).
7 '[Given] a world-wide equalization of wage rates for work of equal low-skill content ... only those in the elite high-skill workforces, particularly those able to partake via multinationals in the international rewards for skilled professionals, managers and technicians, will be able to command high and increasing wages. For the rest, a leveling-down to [the wages] received in the emergent developing nations is not so far-fetched in the long run' (Ashton et al. 1989, quoted in Leys 1990: 121).
8 See in particular Harvey (1982, esp. chs 12 and 13, 1985a, 1985b).
9 Offe's widely cited book *Contradictions of the Welfare State* (1984) was part of this, as was O'Connor's *Fiscal Crisis of the State* (1973), and Crozier's edited collection *The Crisis of Democracy* (1975). Significantly, in his introduction to Offe's book, John Keane, who edited the English-language version, writes: 'Offe does not consider whether the present transnational migration of industrial capital to the peripheral capitalist countries is a direct response by capital to this encroachment upon its power' (1984: 17) and cites approvingly the study of Fröbel et al., *The New International Division of Labour*, which appeared in 1980 and was to prove highly influential.
10 On the other hand, in some cases it would seem that it has also meshed with the goals of particular sectors of the economy. Webber, in a paper that underlines the unity of state and capital that I am emphasizing here, shows how in the case of policy changes in Australia globalization may have resulted less from the desire to reimpose the disciplines of the market and be more a matter of intersectoral struggle: between in that instance manufacturing industry that had been protected and the mining, and agricultural industries, and more recently, retailing which wanted those barriers lifted so as to reduce their own costs of operation, 'It is possible to argue that recent changes in policy are less a matter of deregulation than an element of a process of restructuring, a change in the strategy of growth away from the import-substituting industrialization of the early post-war years and into a policy of internationalization. Economics, by this argument, is not separate from politics. As this example intimates, it is also possible to argue that globalization has been the process whereby a whole host of nations have altered their strategies of growth, that we have created global pressures as much as being their innocent victims' (1998: 35).
11 An argument of diminishing importance with changes in the balance of political forces within the family and the growth of single parent families.
12 This found concrete expression in the Northeast–Midwest Congressional Coalition (Markusen 1987: ch. 8).
13 Admittedly a contradiction in terms, but that was what he called for.
14 Which was primarily one of converting less developed areas into sources of raw materials and markets for the expanding industries of the more developed.
15 It will be noted by some that I have eschewed reference to regulation theory. Given that this has acquired widespread currency in contemporary economic geography this might seem surprising. What is surprising to me, however, is that it has acquired such currency! Brenner and Glick (1991) rebutted its empirical supports

some while back and convincingly. This is not to reject the fundamental premise of regulation theory that capitalist development requires some form of regulation if accumulation is to proceed. It is, however, to reject its discontinuist assumption that regulation can be understood in terms of discrete periods rather than through recognition of continuing transformations in those forms of corporate organization, state intervention, which, through their regulatory force, sustain accumulation. The homogenizing claims of regulation theory – discrete spaces as well as discrete periods – should also be looked at skeptically, as Sayer (1989) has urged.

References

Ashton D, Green F and Hoskins M (1989) 'The Training System of British Capitalism: Changes and Prospects', in P Nolan and F Green (eds), *The Restructuring of the UK Economy*. London; Verso Press. pp. 131–54.

Berry BJL (1967) *Geography of Market Centers and Retail Distribution*. Englewood Cliffs, NJ: Prentice–Hall.

Brenner R (1998) 'The Economics of Global Turbulence', *New Left Review* No. 229, pp. 1–265.

Brenner R and Glick M (1991) 'The Regulation Approach: Theory and History', *New Left Review*, No. 188, 45–119.

Burawoy M (1985) *The Politics of Production*. London: Verso Press.

Crozier M, Huntington SP and Watanuki J (1975) *The Crisis of Democracy*. New York: New York University Press.

Curry L (1972) 'A Spatial Analysis of Gravity Flows', *Regional Studies* 6, 131–47.

Epstein G (1996) 'International Capital Mobility and the Scope for National Economic Management', in R Boyer and D Drache (eds), *States Against Markets*. London and New York: Routledge. Ch. 9.

Fröbel F, Heinrichs J and Kreye O (1980) *The New International Division of Labour*. Cambridge: Cambridge University Press.

Gill SR and Law D (1989) 'Global Hegemony and the Structural Power of Capital', *International Studies Quarterly* 33, 475–99.

Haggett P (1965) *Locational Analysis in Human Geography*. London: Edward Arnold.

Harvey D (1982) *The Limits to Capital*. Chicago: Chicago University Press.

Harvey D (1985a) 'The Place of Urban Politics in the Geography of Uneven Capitalist Development', in *The Urbanization of Capital*. Baltimore, MD: Johns Hopkins University Press. Ch. 6.

Harvey D (1985b) 'The Geopolitics of Capitalism', in D Gregory and J Urry (eds), *Social Relations and Spatial Structures*. London: Macmillan. Ch 7.

Harvey D (1989) 'From Managerialism to Entrepreneurialism: The Transformation in Urban Governance in Late Capitalism', *Geografiska Annaler* 71B, 3–17.

Hirst P and Thompson G (1996) *Globalization in Question*. Cambridge: Polity Press.

Kleinknecht A and ter Wengel J (1998) 'The Myth of Economic Globalisation', *Cambridge Journal of Economics* 22, 637–47.

Leys C (1990) 'Still a Question of Hegemony', *New Left Review* No. 181, 119–28.

Mackinder HJ (1919) *Democratic Ideals and Reality*. London: Constable.

Mann M (1995) 'Sources of Variation in Working Class Movements in Twentieth Century Europe', *New Left Review* No. 212, 14–54.
Mann M (1997) 'Has Globalization Ended the Rise and Rise of the Nation-State?', *Review of International Political Economy* 4 (3), 472–96.
Markusen AR (1987) *Regions*. Totowa, NJ: Rowman and Littlefield.
Massey D (1984) *Spatial Divisions of Labour*. London: Macmillan.
Moellering H and Tobler WR (1972) 'Geographical variances', *Geographical Analysis* 4, 34–50.
O'Connor J (1973) *The Fiscal Crisis of the State*. New York: St. Martin's Press.
Offe C (1984) *Contradictions of the Welfare State* (ed. J Keane). Cambridge, MA: MIT Press.
Piven FF (1995) 'Is It Global Economics or Neo-Laissez-Faire?', *New Left Review* No. 213, 107–15.
Radice HK (1984) 'The National Economy: A Keynesian Myth?', *Capital and Class* 22, 111–40.
Radice HK (2001) 'Understanding Globalization and International Political Economy'. Unpublished paper, University of Leeds.
Savage M (1987) 'Understanding Political Alignments in Contemporary Britain: Do Localities Matter?', *Political Geography Quarterly* 6, 53–76.
Sayer A (1986) 'Industrial Location on a World Scale: The Case of the Semiconductor Industry', in AJ Scott and M Storper (eds), *Production, Work, Territory*. Boston, MA: Allen and Unwin. Ch. 6.
Sayer A (1989) 'Postfordism in Question', *International Journal of Urban and Regional Research* 13 (4), 666–95.
Sayer A (1991) 'Behind the Locality Debate: Deconstructing Geography's Dualisms', *Environment and Planning A*, 23, 283–308.
Scott AJ (1996) 'Regional Motors of the Global Economy', *Futures* 28, 391–411.
Shaw M (1997) 'The State of Globalization: Towards a Theory of State Transformation', *Review of International Political Economy* 4 (3), 497–513.
Storper M (1987) 'The Post-Enlightenment Challenge to Urban Studies', *Environment and Planning D: Society and Space* 5 (4), 418–26.
Streeck M (1992) 'Inclusion and Secession: Questions on the Boundaries of Associational Democracy', *Politics and Society* 20, 513–20.
Webber M (1998) 'Producing Globalization: Apparel and the Australian State', in A Herod, GÓ Tuathail and S Roberts (eds), *An Unruly World? Globalization, Governance and Geography*. London: Routledge. Ch. 6.
Weeks J (1981) *Capital and Exploitation*. London: Edward Arnold.
Weiss L (1997) 'Globalization and the Myth of the Powerless State', *New Left Review* No. 225, 3–27.

10 Time : Space

Mike Crang

Geography has occasionally sought to claim to be a spatial science; sometimes it has sought to define itself through 'areal differentiation', or the synthesis of different factors in specific environs. Indeed, commonsensically, geography has tied its remit to definitions that put 'space' at the centre of geography. And yet if we ask a class of first year students what space is, the most constructive answer tends to be 'the final frontier'. In fact it is probably true that most students arrive at university with up to nine years' study of geography, and yet rarely have they worked through explicit theorizations of space. It is taken as obvious, as self-evident and not really in need of further examination. I mean once we have located things there hardly seems more to be said. Except if you do ask for a little more clarity, as to what space is, suddenly things begin to unravel. Our assurance in using the word space yet our lack of specificity about it reminds me of the discussion of time by St Augustine, Bishop of Hippo, in 397 AD,

> What, then, is time? If no one asks me, I know what it is. If I wish to explain it to him who asks me, I do not know. (*Confessions*, Book 11, Chapter 14)

Both time and space are everyday terms which everyone commonsensically understands and thus are often undefined. This lack of definition is exacerbated by disciplinary divisions since geography has tended to define itself as spatial, and thus tended to regard time as not its province. Thus asking geography students about notions of space produces a variety of answers, whereas if I ask those geography students about time then they draw a blank. If time is considered problematic it is in the sense of finding an appropriate scale to match the process and phenomena studied – just as with spatial categories. The only question is how much of each gets looked at, rather than their actual nature. Recently I shall suggest, and other chapters in this volume show, there has been quite a bit of geographical attention to what space actually is, its relationship with scalar categories such as regions, localities or indeed with notions of place. I want to follow through these issues to unpack the 'space' side of the dichotomy as this chapter progresses. In this way I want to disrupt the binary by illustrating that space is neither self-evident nor self-sufficient but is rather often mutually and problematically defined by and with problematic concepts of time. And it is at this latter point that we have to concede geographers have not developed an extensive engagement.

So both space, with which geography identifies, and time seem to be so obvious as categories as to not need further scrutiny. Except I shall argue that they both carry so much baggage and so many different meanings, that both need careful attention. So in this chapter I suggest, first, that both space and time have multiple facets and definitions. Second, not only are they both complex in themselves but also they tend to be mutually defined (or undefined). Thus definitions of time often lean, explicitly or implicitly, on definitions of space and vice versa. I shall try and show that sometimes this is a process whereby time is likened to space (but rarely the other way round) and paradoxically space is defined as the opposite of time in a classic dualism. These analytic divisions then drive different ontologies and philosophical positions. Thus, for instance, Doreen Massey (1992, 1998) has argued that for radical politics the traditionally important category has been time, which is associated with the possibilities of progress and change. Typically this involved linking the dualism of space and time with another great philosophical dualism – being and becoming. Being is about enduring essences and entities whereas becoming is about process unfolding over time. Space is elided with Being, time with Becoming. The result, to paraphrase many, many reworkings of an essay by Foucault, is that time is seen as fecund and creative while space is seen as passive and inert. And yet over the past decade or two there has been something of an inversion of this, or to use Soja's (1989) sub-title – a reassertion of space in social theory.

Some have argued there is an epochal shift in concerns from time to space, that our concepts of relationships between space and time are being impacted by social changes. Thus we find the very often invoked quote from Michel Foucault that:

> The great obsession of the nineteenth century was, as we know, history: with its themes of development and of suspension, of crisis, and cycle, themes of the ever-accumulating past, with its great preponderance of dead men and the menacing glaciation of the world. [...] The present epoch will perhaps be above all the epoch of space. We are in the epoch of simultaneity: we are in the epoch of juxtaposition, the epoch of the near and far, of the side-by-side, of the dispersed. We are at a moment. I believe, when our experience of the world is less that of a long life developing through time than that of a network that connects points and intersects with its own skein. ('Of Other Spaces' (1967), reprinted in Foucault 1986: 22)

This programmatic statement has come to be used, just as I use it here, as a masthead for an account that sees theory moving to engage with the

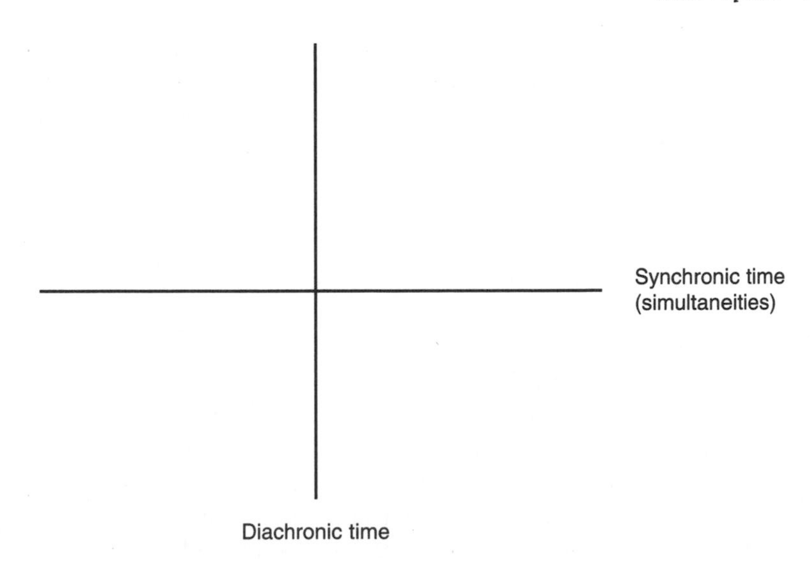

FIGURE 10.1 Synchronic and diachronic time in the built environment (after Tschumi 1999)

spatial. This theoretical shift is argued to parallel social and material changes, in a world of globalized media and trade, where the effects of shifts in one market are felt instantly in another. Thus Fred Jameson depicts a world characterized by simultaneity, producing an aesthetic of pastiche which amounts to the eclectic assemblage of the forms of many eras. Put simply, instead of a linear process of one style succeeding from another, or being seen as superior to it, they coexist at the same time (Jameson 1998). The architect Bernard Tschumi (1999: 170) graphs this simply, labelling space as 'synchronic time' – the coexistence of objects in the same time (see Figure 10.1)

Jameson thus argues what is required is a new way of keeping our bearings – a 'cognitive mapping' to cope with this temporal coexistence. Jameson suggests then that in the current epoch the language of theory, and possibly its ontological concern, shifts from temporal unfolding to spatial occlusion – instead of the consequences of actions being played out over history, the primary issue is one of their rapid diffusion over space; instead of time hiding the consequences, it is space. Or in artistic movements 'the organisation of space has become the primary aesthetic problem of mid-twentieth century culture, as the problem of time (in Bergson, Proust and Joyce) was the primary aesthetic problem of the first decades of this century' (Daniel Bell, cited in Harvey 1989: 201). Figuring the zeitgeist of our age in this manner may be appealing but it clearly overstates any putative shift in concern. As Fred Jameson recently sardonically commented:

> After the end of history, what? No further beginnings being foreseen, it can only be the end of something else. But modernism had already ended some

> time ago and with it, presumably, time itself, as it was widely rumored that space was supposed to replace time in the general ontological scheme of things. At the very least time had become a non-person and people stopped writing about it. (2003: 695)

He goes on to pose a difficult question which is whether some of the theoretical concepts I am going to develop towards the end of the chapter can successfully patch together a dualism when it too often seems that 'Time and Space are at war in a Homeric combat' (p. 698). So, rather than a disciplinary celebration that social theory seems to have noticed geography, we need to unpack how space and time interact.

So my first step is to unpack a range of types of space and then outline a similar range of ideas of time. Thus at the very least we can suggest there is no simple binary between two terms, but relationships between different sets of terms, and thus often different relationships between specific different elements. What I then hope to suggest is that many of these specific and varying definitions are subtended by specific and varying dualisms – and to illustrate this I want to follow through two ways spatial models have been used to understand time. In the first, I will examine how one of the ontologies of space identified in the following section – abstract space – has been seen as, at best, informing problematic understandings of time and, at worst, rather obfuscating the whole notion of temporality. The relationship of this type of space to time has been used to illustrate the problems of spatializing time, and lies at the heart of how space has often been regarded as secondary to history. I want to suggest that at the least we see this as a problem not maybe inherent in spatial notions, but only in using those particular spatial notions of abstract space. The second example I then want to use takes a looser epistemology about space, that of landscape as convergence, to indicate a multiplicity of temporalities interacting. Here space is used to free up and pluralize what 'time' entails. My final move then is to look at categories that seek to link space and time – and here I want to focus upon a concept taken from the literary critic Mikhail Bakhtin – the chronotope.

Multiple spaces

This section recaps some of the issues in Chapter 5 (by John Agnew) to highlight senses of space that have played an important role in relation to notions of time. If we start with the most obvious sense of space, it is that of location. And what do we mean at bottom by 'located'? A definition often might start, and all too quickly all too often end, with co-ordinates, x and y, east–west and north–south as a form of Newtonian space. It tends to imply a sense of space as length, area or volume, that is infinitely divisible into units (x_1, x_2, x_3, ... x_n, y_1, y_2, y_3 ... y_n, and thus locations combining these). The implication then is that space is about difference in location, not type. That is, we tend to assume two things. First, that objects are independent of location – a house at one end of a street might be the same as one at the other, and the only difference between them is position. This is not to say there is no difference, one need only think of the three key factors in house prices ('Location, location, location') to see that spatial position matters. But the type of difference is enumerative or quantitative. We can see this notion of space taken to its logical conclusion in the assumptions behind the spatial models of von Thunen or Christaller – an isotropic plain, where space is homogenous and quantifiable, where all other differences are stripped out. Such a concept of space has real consequences. Perhaps the most famous example is the mapping of the west of the United States where, beyond Route 277 in Ohio, land was marked out for potential homesteads, settlements and townships from great sweeping meridians charted across the land with such regularity that it was compared to graph paper (Linklater 2002: 178). It drained out all substantive content from space and replaced it with empty, exchangeable units of measurement. What this facilitated was the rapid commodification of land – with standard plot sizes and multiples of plot sizes, identified by unique location in the grid. Second, then, this version of space sees territory as divisible and multiplicable. In other words, since the only difference between places is location, they can be ever more finely sub-divided into smaller components, or indeed added together to form larger units. The only change is quantity. As Lefebvre put it: 'This is a space, therefore, that is *homogeneous yet at the same time broken up into fragments*' (Lefebvre 1991: 342, emphasis in original).

There are numerous alternative ways of conceptualizing space from which I am going to select three that highlight themes that emerge later. So we might first make an argument that this empty space is actually too solid, that it gives meaning to space before content. A relational, Leibnizian view of space instead sees it as only defined by the objects within it and their relationships one to the other. For instance, a feudal society where a land grant might be measured in land sufficient to graze a herd of pigs (to take one form of measurement used in England's eleventh-century Domesday Book) has a very different notion of space than one where standard units of length and width define ownership. The differing sense of space of different epochs is not just a story about accuracy of measurement, it is about differing societies' relationship to space; as Lefebvre would say, not just a matter of social relations and contradictions *in* space but *of* space (Lefebvre 1991: 334). To put it another way, societies do not just occur in a pre-given space where the only question is how much of it they occupy, but they actually create space. Thus even if we think of the empty space created in the surveying and dividing of the American West, this was a space produced in order to facilitate and enable its colonization (and balanced the interests of state, farmers and land speculators). Its 'emptiness' and the lack of substantive meaning for any given place is created, it is a 'semantic void that abolishes former meanings' (Lefebvre 1991: 307).

This abstract space can be counterposed to 'inhabited' personalized meaning given to places. As Michel de Certeau put it, confusingly inverting the terms from how geographers generally use them (see Chapter 5), space is inhabited place (de Certeau 1984, 1985). To coin a phrase, houses differ by more than location when one of them becomes our home. This division of place and space has been around for a long time, and we can certainly trace it as far back as Plato and Aristole, with notions of space as *kenon*, that is a void, or as *chora*, variously taken as meaning formless or receptacle or container and *topos* which is the known, bounded and inhabited place (Grosz 1995). It is possible to read *chora* then as an unstable term, as neither place nor space in the work of Derrida for instance (Eisenman, 1997: 134). However, *chora* has most often been read along with *kenon* as emphasizing space as a raw clay waiting to be moulded, or perhaps more accurately a vessel waiting to be filled. I want to start thinking more precisely about *topos* too. We might think of sacred spaces, as being what Lefebvre (1991) would call an absolute space. It is not a matter of these being portable, but rather there is a specific investment in one location. These sites are precisely not interchangeable. This is the origin for the idea of *genius loci*, the spirit of place, but more interestingly this is generally now taken to suggest that special quality of place, created through long-term attachment and the convergence of many factors – the daily rhythms, personal histories and secular and/or religious rituals. The emphasis upon the convergence suggests this is about fusion into a new whole. It is not then about 'extensive boundaries', about defining place through the limits of spatial extension, but rather it is an intensive threshold, where there is an internal transformation, like a phase shift (de Landa 1998, 1999).

If we think back to our first list of binaries, what is apparent is that notions of 'intensive threshold' or 'habituation' suggest place Becomes – it is not simply Being. It is made and remade, and, for that matter, undone. We might approach this then through a vocabulary that sees space as action not location. One way into this is through notions of dwelling which, following Heidegger, we might see as the activity of being-in-the-world. Certainly we can begin to push the affective dimensions of space – its emotional resonance, such as senses of security, and its specificity. Thus Gaston Bachelard (1964) identifies spatial archetypes of security such as the den but also further territorial binaries of inside and outside. But more strongly, there is the sense that this is not space containing objects but space that is created through actions and thus we might better talk of spacing. In Heidegger's analysis of a Greek temple he suggests it is not added to a given place but that rather 'the building precedes its site' (Wigley 1993: 61). That is, it is the temple that creates the sense of sacred space, it creates the ground for its people (Elden 2001: 66). Taking the example of Hölderlin's poetry on the great German rivers, Heidegger argues the rivers are not symbols of various places but create those places, they make the lands associated with them not just in a geomorphological sense but in the sense of a dwelt territory, and thus the poems are performing placing or are 'platial' as opposed to spatial (Elden 2001: 36).

The reason for delving into Heidegger is that first, as we shall see, he offers fruitful connections into issues of time – which is his main concern. Second, and of more immediate interest here, he moves us from seeing abstract space as 'objective' and place as 'subjective'. He pushes against this dualism by insisting on the objectivity of dwelling. He argues that all understanding comes from Being in place, by being situated. His focus is neither subject nor object but situation (Jameson 1999: 213). In other words, there is no unplaced knowledge, no transcendent viewing point and no unplaced transcendent subject. While Heidegger's position can be dangerously conservative, he critiques abstract knowledge for depending upon modes of representation which distance us from an experiential engagement. This critique finds echoes with Lefebvre's well-known triptych of representations of space, representational space and spatial practice (1991: 40–6) – which can be glossed as abstract ideas of space, affective and finally lived space respectively – which has provided the framework for this section. We move from ideological notions of space, that clearly play a role in shaping societies, their structures and activities, through to the sense of the meaning and emotional resonance of places before finally thinking about the habitation of those places – not their representation through mediating schema but their direct interrelationship with people's activities and identities. It is revealing to look at how some of the same issues of multiple meanings are replicated in discussions over time.

Multiple times

If space is often taken as a common sense element and a shared datum which can anchor analyses, the same can be said of time with added force, since there is the added force of direction – time often being defined as different from space in that it is irreversible. It is this sense of flow that clearly underpins the notion of time as Becoming and space as Being, time as action and space as context. However, it is also the case that time is by no means self-evident. In fact to start with lived time we might say it is more dominated by cycles than linear flows. 'Everyday life is above all a temporal term. As such it conveys the fact of repetition; it refers not to the singular or unique but to that which happens "day after day"' (Felski 2000: 18). This is not a time of developmental logic and conscious planning. Cullen suggests the imperatives of routines mean that in quantitative terms, deliberative choices

> are swamped by a dominant pattern of repetition and routine. We spend very little time each day either deliberating some future action or executing a previously deliberated one. Most of our time is devoted to living out a fairly sophisticated pattern of well ordered and neatly integrated routine. (Cullen 1978: 31)

In other words, our daily lives have a temporality that is often not the linear flow of time's arrow but is composed of cycles. Getting up, going to work, eating meals, weekdays and weekends all occur with stupendous regularity when we look at Western societies. On longer scales there are the cycles of birthdays and festivals. In other words ritual times, be they religious, personal, commercial or mixtures of all of these, are often cyclical. Hall (1983) takes the example of the Quiche Maya who traditionally had both sacred and civil calendars of each of which had different numbers of different-length months which interlocked to only repeat every 52 years. As opposed to Anglo-European calendars, where the cycle of routine overlays largely undifferentiated days, here each has a proper name in the sacred calendar (1983: 81). If we look at the Dreamtime of Australian Aboriginals we find it is one of continued recurrence and relevance – it is a mythic time marked by being out of our current time, that is not past but is continually present (Perkins 1998). And from the sacred we should not forget the opposite – profane times, such as the time of carnival or other leisure times where the normal order is overturned. So cyclical time can be important in a number of registers. Hall concludes we can identify at least eight clusters of types of time divided around key oppositions: thus some might be interpreted as physical (for example, ageing, the seasons) while other issues are cultural (such as religion); some times might be defined as individual or collective, some are 'exogenous' (where they appear objective) and others are contextual (dependent on the beholder), to replay the objective and subjective dichotomy. Putting these together we end up with (1) sacred, (2) profane, (3) micro scale, (4) synchronized, (5) personal, (6) biological, (7) physical and (8) metaphysical types of times (Hall 1983: 17).

Just to add to the complexity, we might chart the changing historical balance of linear and cyclical time. Analysts have often pointed to the association of women with reproductive labour, and suggested this has had a more cyclical character – whereas men accessing the public realm had access to a public time of historical progress. Historically, the same pattern inflects class experience, with linear time being associated with a progressive self-narrative of self-actualization or accomplishment that we find emerging in the early modern period among the bourgeoisie – buttressed by congeries of new technologies. So here we find shifts from the book of hours, and prayer books suggesting devotional activities for each hour of each day, to personal reflective diaries that spin the self, and self-understanding, into a temporal narrative. It is often argued this narrative is connected to a linear time that slowly diffuses through society (Maynes 1989, 1992). It is common in studies of literary and textualized versions of time to suggest that the ability to grasp oneself in time is indeed to sense one's life as narrative (Currie 1999: 45). The self then becomes a unity through having a narrative development between beginning and ending. Alternate accounts look to the rise of capitalism with the time of the ledger book triumphing over the medieval time, full of chimes ringing out calls to church or mosque (Le Goff 1980). If we have seen abstract space as making land a commodity, then we might follow Lewis Mumford in defining the clock as the key machine of industrial capitalism (Nowotny 1994: 47). Each minute of the day becomes calculable and measurable – to be bought and sold between capital and labour. We have to be cautious about opposing cyclical and linear time, where for instance Hegel, and indeed Marx following him, identified Hindu India with cyclical time and thus with a lack of progress, and the British colonizers with forward-looking dynamic modernity (Spivak 1991). More nuanced accounts of religious temporalities look to monasteries themselves as incubators of new technologies of time as they invented orders and sequences of devotional activities dependent upon their timing through the day and year. Indeed religion itself has pushed the notion of the narrative self, as in Augustine, and the sense of time as a something to be spent wisely (Fenn 2001).

Even when we begin to think then of time as flow we immediately encounter paradoxes. Let us return to Augustine and his great narrative recomposition of the self in the *Confessions*. In Book 11, where he explicitly thinks through the implications of this for time, he first formulates a notion of the disappearing present – no sooner thought than it has gone. That is, the present is not so much a day or an hour or even a second as the fine boundary line between the future and the past, which I will suggest we should think of not so much as itself moving forwards as being the line through which the future flows into the past. As Henri Bergson put it, the present is not so much '*that which is*, [but] ... simply *what is being made*. Nothing *is* less than the present moment, if you understand by that the indivisible limit that divides the past from the future. When we think this present is going to be, it exists not yet, and when we think of it as existing, it is already past' (1991: 149–50). The present as the only realm in which we act thus appears to shrink away while the past and future quite clearly have different ontological characters – the one not yet being, the other having ceased to be. Grappling with this same problem, Augustine reversed this conclusion – producing what we might call the 'big now' (Ricoeur 1988: 30) instead of the disappearing present:

> But even now it is manifest and clear that there are neither times future nor times past. Thus it is not properly said that there are three times, past, present, and future. Perhaps it might be said rightly that there are three times: a time present of things past; a time present of things present; and a time present of things future. For these three do coexist somehow in the soul, for otherwise I could not see them. The time present of things past is memory; the time present of things present is direct experience; the time present of things future is expectation. (Book 11, Chapter 20)

In this way Augustine moves to ground temporality in human experience rather than an external measure. We reach forward to grasp the future while carrying with us our past – and thus we make both elements present. The term Augustine uses is '*distentio animi*' for the stretching of the spirit to include our expectations and our memories. The mind expects, it is attentive and it remembers (Alliez 1996: 131). Our own experience will tell us that this is not a uniform sense of time as there are occasions when time seems to crawl towards one anticipated event and hurtle away from another fondly remembered one (Flaherty 1999).

No moment is then fully self-contained, or as we shall see, the present is then precisely not presence. As Augustine formulated it in his Chapter 14, 'If, then, time present – if it be time – comes into existence only because it passes into time past, how can we say that even this is, since the cause of its being is that it will cease to be? Thus, can we not truly say that time *is* only as it tends toward nonbeing?' Or to restate the significance of this, 'The fragile hold of the present on reality ... is itself encroached upon by the surrounding voracious non-existence of past and future' (Lloyd 1993: 22). The insight of the extended present has been carried through into a range of temporal ontologies. In Husserl's phenomenology every instant is marked by protention and retention, carrying with it the traces of the past and the seeds of the future. This approach perhaps reaches its apogee with Martin Heidegger, who looks at temporal Being as matter of three *ekstases* or modes of time: a being-alongside, that is simultaneity, a being-towards-death, that recognizes the force of time's arrow for all humans, and 'thrownness', that is we find ourselves cast into the world in situations neither of our making nor choosing. What is more, the depth and shape of these modes is shaped by our structure of care and concern towards the world – in other words, the time frame that we bring to bear from past and future varies according to the sort of tasks we are undertaking. The fundamental insight Heidegger then provides is that life and the subject are temporal. We do not develop some notion of the human subject and then insert it into grids of space and time. Rather the subject is formed through structures of temporality and spatiality. This sense then of a tripartite time where the present is so insubstantial and always slides past I want to contrast with the dominance of a form of abstract time, that sees the present as a definable instant.

Time as space (part 1)

Very often time is deployed as though it were a series of salami slices or beads upon a string, a series of 'nows' that follow one after another in a sequence. The model for this is clearly that of abstract space, where to our spatial coordinates we add temporal ones in an infinite and empty series (t_1, t_2, t_3, ... t_n). As Hall puts it, taken-for-granted Anglo-European 'time is an empty container waiting to be filled' (Hall 1983: 84). Perhaps the classic example of this treating of space and time as an external container is time-geography. Here time is added as another axis to the conventional map in order to produce an action space, a container of possibilities. An intriguing pattern then emerges of activity prisms (the scope for movement between fixed points in space–time) which lead to coupling constraints (how different people's activities can be coordinated) in a delicate and fascinating choreography of people's paths through space–time. Thus, while the great pioneer of time-geography, Thorsten Hågerstrand, sympathized with some phenomenology of time, he argued resolutely that 'external' and objective time were the key dimensions (1982: 324). In this vision then people may move quicker or slower, but they share a set of time–space dimensions. However, Grosz argues this means that:

> Even today the equation of temporal relations with the continuum of numbers assumes that time is isomorphic with space, and that space and time exist as a continuum, a unified totality. Time is capable of representation only through its subordination to space and spatial models. (1995: 95)

However, in utilizing precisely the concepts of abstract space to depict time it presents very much a sequence of moments. For all that it plots the paths and trajectories of actors, we might say it succumbs to a cinematic illusion about time. Thus, if we turn back to the early twentieth-century philosopher Henri Bergson, we find him arguing that this approach misses precisely the sense of flow in time as it

> masks the perception of real movement … your succession of points are at bottom, only so many imaginary halts. You substitute the path for the journey, and because the journey is subtended by the path, you think the two should coincide. But how should progress coincide with a thing, a movement with an immobility? (1991: 189–90)

In other words, we lose precisely that sense of expectation and retention, or being towards the future and thrownness. Or to take the gloss on his ideas from Gilles Deleuze, time and space are qualitatively, ontologically different and thus using one to describe the other creates a compound concept that is inherently chaotic because

> movement is distinct from the space covered. Space covered is past, movement is the present, the act of covering. The space covered is divisible, indeed infinitely divisible, whilst movement is indivisible, or cannot be divided without changing qualitatively each time it is divided. This already presupposes a more complex idea: the spaces covered all belong to a single, identical, homogeneous space while the movements are heterogeneous, irreducible among themselves. (1986: 1)

I think this analysis is correct as far as it goes. Time is not reducible to this sense of space. But these are moves that define time as vibrant difference and space as repetition – by opposing time as flow, or in Bergson's terms '*durée*', to space as coordinates. It is an important critique of a dominant form of a 'universal time [that] appears to be no more than a hypothetical projection, a time of generalized equivalence, a "flattened" capitalistic time' (Guattari 1992: 15). The critique's fundamental point then is to see time as about phase shifts, shifts of kind rather than just a movement in temporal location. The sense of time as just a series of instants and points denies this sense of qualitative difference – producing what Castoriadis calls 'identitiary time' since all the instants are ontologically identical:

> In this identitiary time exists the identitiary present and, reciprocally, identitiary time is but the innumerable (and numbered) repetition of identitiary presents, always identical as such and different only by their place. (1987: 201)

This forms the basis of what he calls 'public time'. The sense of time here then we might label as 'chrono-time', to signify the procession of empty sequences. Far from being the 'real time', or objective time, if past and future are implicated in each other then moments are not discrete objects but have a temporal unfolding and this is a post hoc representation of duration (Lloyd 1993: 98). Instead of our stories being fictional spins over the unyielding sequence of time, it suggest that '[t]he reality of our temporal experience is that it is organised and structured; it is the "mere sequence" that has turned out to be fictional' (Carr 1986: 25). If we follow the idea of a phase shift between the past, future and present, as an intensive boundary not an extensive one, we could interpret both past and future, though inaccessible, as real, but as a virtual order rather than an actual one. We have then a more complex pattern of oppositions – the virtual (that which might be) is the antonym of the actual (that which is), but both are ontologically real. This shift in understanding suggest that while space preserves elements, time devours them:

> Duration is the continuous progress of the past which gnaws into the future and swells as it advances. And as the past grows without ceasing, so also there are no limits to its preservation. Memory ... is not a faculty of putting away recollections in a drawer, or of inscribing them in a register ... In its entirety, probably, it [the past] follows us at every instant ... leaning over the present which is about to join it, pressing against the portal of consciousness that would fain leave it outside. (Bergson 1991: 7)

Here Bergson suggests that each instant of the present brings with it a vast virtual order (Figure 10.2) of memories stretching out from around our present action and called forth by our orientation in the world. Depending on the issue upon which we are focused, a different range of these memories is brought to bear – more narrowly or widely focused. However, they all still suffuse each specific instant. He argues for the reality and persistence of the

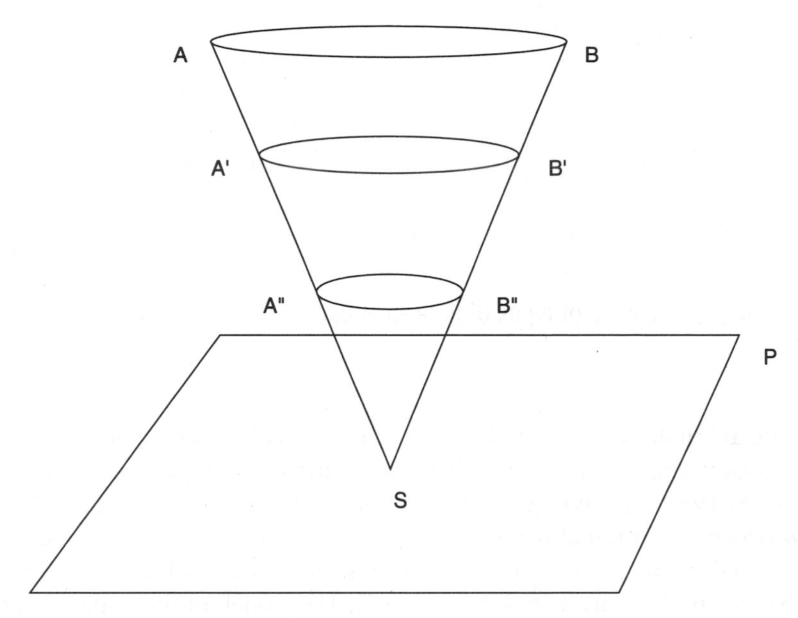

FIGURE 10.2 The plane of the present and the distention of memory (redrawn from Bergson 1991: 162 (1991: 210) reproduced with permission)

past *without* committing it to being the same kind of thing as space. The present and past coexist in a virtual order:

> We have great difficulty in understanding a survival of the past in itself because we believe that the past is no longer, that it has ceased to be. We have thus confused Being with being-present. Nevertheless, the present *is not*; rather it is pure becoming, always outside itself. It *is* not, but it acts. Its proper element is not being but the active or useful. The past, on the other hand, has ceased to act or be useful. But it has not ceased to be. Useless, inactive, impassive it IS, in the full sense of the word: it is identical with being in itself. (Deleuze 1991: 55; original emphasis)

In a reverse of how we often think of time, the past does not recede but 'literally moves towards the present' and exerts a pressure to be admitted (1991: 70).

This seems an important step forwards, yet Bergsonian *durée*, while not being a linear sequence can give the impression of being a smoothly flowing time. We need to emphasize the notion of attention to the world, as in Heidegger, that orchestrates the recall of memories and the production of possible futures. This sense of time is perhaps best summed up as opportunity, not just about action but acting at the right time. To give this sense of time a name we might use 'the Greek *kairos*: what we might call the temporal opportunities of everyday life' (Maffesoli 1998: 108, 110). This is the sense of uniqueness of each moment, of its specific and irreproducible nature.

		Time	
		Chronos	*Kairos*
Space	*Chora*		
	Topos		

FIGURE 10.3 A schema of types of time–spaces

To use an analogy then, 'If chronological time is like worldwide suburbia, kairological time is the *genius loci*, the spirit of that particular moment' (Griffiths 1999: 22). We might thus consider whether for each place with its *genius loci* there is not also a *genius tempori*. We might then think about these two types of time – chronological and kairological and their interaction with space as *chora* and *topos* (Figure 10.3). The models of time–space in time-geography tend to be chrono-chora, what Deleuze terms 'anywhere-whenevers' (1989), of abstractly identical time–space units, and we can see Bergsonian influences in formulating a kairo-chora relationship. What is less often discussed are chrono-topos and kairo-topos senses of time–space (Rämö 1999). It is to these latter two that I wish to turn in the next sections.

These different senses of time have a dialectical relationship in the arts. Just as 1913 saw the Eiffel Tower broadcast the hour 'pips', abolishing local times in France, it also saw the publication of Proust's *À la recherche du temps perdu* and 'time, which in one sense had never been so public, so monolithic, had in another sense, with Proust, never been so private, so unique, so local to the psyche's own hour' where it was 'all madeleines and murmuring memories of the time of the mind' (Griffiths 1999: 19). Far from being linear, it reflects the concern and attention of the mind, where one day is allocated 287 pages while another whole year does not even merit a mention, so much that Proust concludes his mammoth account that it is through time that humans occupy 'a place, a very considerable place compared with the restricted one which is allotted to them in space, a place on the contrary prolonged past measure – for simultaneously, like giants plunged into the years, they touch epochs that are immensely far apart, separated by the slow accretion of many, many days – in the dimension of Time' (1999: 19). This sense of distension and connection then meant Proust had to deny he had written the literary version of Bergson's theory, a '*romans bergsonien*' (Gross 1985: 376). He gives a different inflection than flow and narrative extending and unifying a subject. 'Proust's whole concept of memory was founded on the notion that we are not continuous but altogether discontinuous selves' (Gross 1985: 378). Whereas the phenomenological accounts point to two equal streams of time and consciousness in a perfect parallelism, here we have time disrupting the subject where, after Rousseau, 'Things always happen too early, understanding always comes too late' (Bielik-Robson 2000: 72). Instead of a simply ever-expanding virtual cone of memory, 'The eternity which Proust opens to view is convoluted time, not boundless time. His true interest is in the passage of time in its most real – that is space-bound – form' (Benjamin 1973: 206). While 'Bergson denounces and rejects the metamorphosis of time into space, Proust not only accommodates himself to it but installs himself in it, carries it to extremes and makes it finally one of the principles of his art' (Poulet 1977: 4). But in Proust space is figured in a rather different way, acting to bound and contain an archipelago of incidents, or as Poulet (1977: 90) puts it, forming a dispersed set of 'closed vases', that are left by the withdrawal of life. Space is far from homogeneous or inert.

Time as Space (part 2)

Let us then think about time using some different conceptions of space. I want to start by thinking of space as '*topos*' and using that as a model for time. This approach underlies the recent work of Barbara Adam and her concept of 'timescape' (Adam 1990, 1995). The term is a deliberate utilization of the notion of temporal landscape to talk about how events and actions are located in time. Using landscape allows her to evoke the multiple dimensions of time, and the way they interact to form particular constellations. She outlines five temporal dimensions (Figure 10.4). First there is 'temporality' as duration or brevity, in other words how long a specific event or action lasts. Second is the 'time frame' of how actions relate to each other – whether they occur simultaneously or in a sequence. Third, is the 'time point' by which she points to the frequency and sequence of the action itself – is it intended to recur, to be cyclical, if so how frequently? Fourth is 'tempo', which is about effects and connections, with causality and consequences worked out over time. Fifth is the notion of 'timing', like kairological time, the ability of people to take opportunities or indeed to miss them, or to find them precluded.

The multiple aspects of any given event thus expand beyond simply future–present–past. Adam highlights, for instance, the dimension of synchronizing action – where things must occur in the right sequence and alongside each other as another way of understanding the multiple linkages of events in time rather than just as sequence or linear flow. Adam notes that despite the hegemony of abstract clock time, these other senses of temporality continue so that, for instance, 'embodied time is lived and experienced alongside, despite of, and in conflict with the culturally constituted social relations of time' (Adam 2003: 61). She suggests we can see historical patterns of the five Cs. The creation of time to human design (C1), through technologies like clocks and diaries, but also novels and narratives, then the

Timescape = Time, Space and Matter		
Temporality	Past	Duration ⇔ Instantaneity
Time frame	Present	Sequence ⇔ Simultaneity
Time point	Future	Repetition ⇔ Rhythm ⇔ Beat
Tempo		Cause ⇔ Effect ⇔ Time lag
Timing		Reason ⇔ Action ⇔ Symptom

FIGURE 10.4 Elements of the 'timescape' (after Adam 2002)

commodification of time (C2) as it becomes the criterion of pay and the measure of productivity, the compression of time (C3), as we increasingly look at real-time networks where effects at a distance occur (nearly) simultaneously, and thence the control of time (C4) and finally the colonization of time (C5), both in the sense of the intensification of routines into a 24/7 society but also our ability to discount costs into the future. Using timescape allows us to register that all of these elements are in changing patterns, with alignments and conflicts between different elements. It is just too glib to talk about the dominance of space over time, when instead there are much more complex patterns of different sorts of space and time interacting.

> Cause and effect, linearity, spatiality, invariability, stability, clarity and precision are *not* being replaced but have alongside and superimposed contrasting temporal principles such as instantaneity, simultaneity, networked connections, ephemerality, volatility, uncertainty as well as temporal multiplicity and complexity. Emerging alternative and contradictory temporal principles constitute today a lived reality for a significant number of people across the world. (Adam 2003: 74)

Alternately, we might think of the kairo-topos or what, slightly confusingly for the terms I am using, Mikhail Bakhtin called a chronotope (Holquist 1990; Kneale and Holloway 2000). That is the unity of a specific sense of time with a specific kind of space. So I want to pick up on a writer who played with the relationship of space and time and their representation, and think of James Joyce's *Ulysses*.

This text remains one of the great works of modernism and stream of consciousness style. Ostensibly the story of the peregrinations of one salesman for one day, written as though recording each thought and action as they connect to another, it plays with the conventions of time, space and language. It was a shocking new work, not just for scatological and sexual references, but its linguistic novelty – reputedly it contains the longest unpunctuated sentence in English literature – while it also plays with notions of unfolding plot and time. The structure of the work is the encounters of Leopold Bloom over one day – from waking to a late and rather drunken somnolence. These frankly banal wanderings are sectioned and chaptered by headings lifted from Homer's *Odyssey*. Clearly a set of parallels is being drawn, and for our purposes here we can focus in on the particular aspects highlighted by Umberto Eco – where he suggests Joyce presents a 'chaosmography' (Eco 1989). Eco highlights the paradox of order and chaos, suggesting that the orderly cosmology of the *Odyssey* is deliberately set against the chaotic flux of the modern city. Thus, through a series of inversions and paradoxes the book highlights both continuities and rifts. Most obviously, the text appeals to a mythic time, but then sets heroic Odysseus, one of the first human narrative heroes who struggles to shape his own destiny, against the bathetic figure of Bloom striving to stay afloat in modern society. It is never entirely clear 'whether this parallel [Homeric] plot is an ironic, mocking memory of a heroic past world that emphasizes the emptiness of modern life, or whether it provides a source of enrichment, a promise of wholeness and atonement to come' and how we reconcile the tension 'of an irretrievable past and a paralyzed present' (Rickard 1999: 14, 82).

So, the first temporality is that of the unmoving and eternal that embraces and swallows contemporary narratives. However, the time of the original Odyssey is that of a life journey: departing home, prevailing through travails and travels in a spatial story stretching over 20 years. Ulysses takes the departure from home, though in this case it echoes the three lost Trojan stories, and far from the constant wife maintaining the masculine home, it depicts inconstancy and insecurity for the hero. It also compresses all the wanderings and magical far-off places into a single day in the city – time–space compression indeed. It marks out this hyper-intense day through the flood of intertextual referents that Bloom negotiates – with news bills and papers bringing the world to him. The effect is one of chaos, not just as spatially distant events crowd in to the city, but also where the notion of building a narrative life project, a coherent self through binding past and future dissolves. The book thus appears to collapse time into simultaneities rather than offer temporal development (Tschumi 1999: 170). The single day could be any day in a cycle of repetition, it is unchanging and yet unique. There is no resolution and Joyce leaves us with an unfinished event not gathered into a narrative whole (Schleiffer 2000: 78). So there is fragmentation and yet an overarching frame of reference. The intertextual effects go beyond the Homeric backcloth, since Joyce's method of composing involved an endless burrowing through newspaper reports and snippets about the city and society, and the final text is almost a collage of unsigned quotes and sources. The effect is to decentre the subject and the present which is seen then not as a self-present moment but one that is endlessly responding to the baggage of history – as when Stephen Daedalus sums up the Anglo-Irish conflict by saying 'apparently history is to blame'. More exactly for the sense of the 'big now' it suggests this is not an interior story but a dialogue with the world which reflects 'that each one who enters imagines himself to be the first to enter whereas he is always the last term of a preceding series even if the first term of a succeeding one, each imagining himself to be the first, last, only and alone, whereas he is neither first nor last nor only nor alone in a series originating in and repeated to infinity' (Joyce, *Ulysses* (1961: 731), cited in Schleiffer 2000: 149).

The fragmented self is articulated through an urban space where Stephen Daedalus and Leopold Bloom 'do not go, like Proust's Marcel, in search of lost time: memory is coextensive with their perceptions, manifesting itself in a thousand elusive forms' (Rickard 1999: 129). The city carries the proleptic force of memory as much as any agent and it interjects involuntary memories. It is not then the individual suturing a trajectory over time, but also the city interjecting and disrupting that account. Here 'space functions in fiction through and as temporality, as a narrative event or events' where it represents a network of relationships be they unfolded or not, or in Raymond Williams' words, the 'forces of action have become internal, and in a way there is no longer a city, there is only a man walking through not the history of the city but the loss of city' (Johnson 2000: 200). The past pushes its way into the protagonists' times not as in the sense of Bergson's structures of attention or Heidegger's care, but through the spaces and arrangement of the city. The minute reconstruction of the city in the text is not about totalizing command of the urban scene. So though Joyce wrote to his literary agent Frank Budgen that 'I want to give a picture of Dublin so complete that if the city one day suddenly disappeared from the earth it could be reconstructed out of my book' (Johnson 2000: 199), he presents a city rising up in fragments, not laid out in either historical or spatial order: 'Other novelists are ... much more likely to present the city in reconstructable form. Joyce offers no architectural information, only places to bump elbows, or to lean them, to see out the corner of an eye, to recognize by a familiar smell. The city rises in bits, not in masses' (Ellman, cited in Rickard 1999: 142). This urban scene then works by bringing multiples spaces and times together into a paradoxical relationship.

Concluding remarks

The relationship between time and space I have tried to suggest is complicated by a number of factors. First, the very commonsensical facticity of the two has often meant they are not examined. However, the second issue is that they have often fed from each other in binary oppositions. Third, I have tried to show the binary pairs to be of one specific kind of time against one specific kind of space. The first move to disrupt the binary opposition is thus to recognize the plurality of forms of both space and time. I have in the last two examples tried to suggest how some less common combinations of notions of space and time can produce different understandings. Later in that celebrated and much quoted passage by Foucault on the epoch of space he too says 'space itself has a history in Western experience, and it is not possible to disregard the fatal intersection of time with space'.

These last two examples suggest certainly, as Heidegger put it, 'Time is not a thing' (Schleiffer 2000). It is not simply the rate or duration of events occurring in time but the shaping of the temporal framework in which those events occur. Furthermore, our understanding of that time cannot stand outside time or space. As Derrida suggests 'In a sense, it is always too late to talk about time' (in Jameson 2003: 697) because we and our ideas are always in its flow. We might follow Jameson to suggest we are perhaps always in the wrong place to speak of space. What seems to happen in many theories is that one of space or time is held constant, thus producing rather distorted visions of time–space. Thus, when time is conceived as difference it tends to be through a foil of space conceived as repetition of the same. When space appears as the preserver of past action, time is the destroyer. My aim in this chapter then has been fourfold. At the most basic level, it has been to illustrate that the meaning, and possibly the substance, of both time and space are plural. Second, feeding from this, any binaries tend to be using the characteristics of one definition of space or time against which to define the other term, and any such definition is thus hopelessly over-determined. Third, the two terms cannot really be separated so all action occurs in time–spaces. And finally, this inseparability is not just a matter of bolting two conceptually discrete elements together, but rather that the two are not separable conceptually. Or perhaps, to say that separating them into just two terms is obfuscatory. I have tried to illustrate how geography has related time and space through the heuristic grid of *chronos/kairos* and *chora/topos*, with chrono-chora dominant. It is clear that simple binaries of time and space do not hold, but rather unravel and spawn yet more divisions. In our rethinkings of space and place I suggest we cannot simply ignore temporality and add it later. Rather geographers need to think through time–spaces of various sorts and how they impact on their work.

FURTHER READING

Adam, B. (1990) *Time and Social Theory*. Cambridge: Polity Press

Bachelard, G. (1964) *The Poetics of Space*. New York: Orion Press

Davidson, C. (ed.) (1999) *Anytime*. Cambridge, MA, MIT Press

Elden, S. (2001) *Mapping the Present: Heidegger, Foucault and the Project of a Spatial History*. London: Continuum

Goff Le, J. (1980) *Time and Culture in the Middle Ages.* Chicago, IL: University of Chicago Press

Grosz, E. (1995) *Space, Time and Perversion: Essays on the Politics of Bodies.* London: Routledge

Massey, D. (1992) 'Politics and space/time', *New Left Review* 196: 65–84

May, J. and Thrift, N. (eds) (2001) *Timespace.* London: Routledge

Rämö, H. (1999) 'An Aristotelian human time–space manifold: from *chronochora* to *kairotopos*', *Time & Society* 8 (2): 309–28

References

Adam, B. (1990) *Time and Social Theory*. Cambridge: Polity Press.
Adam, B. (1995) *Timewatch*. Cambridge: Polity Press.
Adam, B. (2002) 'Management in the Context of Globalized Time: Problems and Creative Opportunities'. Dynamic Time and Creative Inquiry in Organizational Change Conference, Boston.
Adam, B. (2003) 'Reflexive Modernization Temporalized', *Theory Culture and Society*, 20 (2): 59–78.
Alliez, E. (1996) *Capital Times*. Minneapolis: Minnesota University Press.
Bachelard, G. (1964) *The Poetics of Space*. New York: Orion Press.
Benjamin, W. (1973) 'The Image of Proust', in *Illuminations*. London: Fontana. pp. 196–210.
Bergson, H. (1991) *Matter and Memory*. New York: Zone Books.
Bielik-Robson, A. (2000) 'Bad Timing: The Subject as a Work of Time', *Angelaki* 5 (3): 71–91.
Carr, D. (1986) *Time, Narrative and History*. Bloomington, IN: Indiana University Press.
Castoriadis, C. (1987) *The Imaginary Institution of Society*. Cambridge: Polity Press.
de Certeau, M. (1984) *The Practice of Everyday Life*. Berkeley, CA: California University Press.
de Certeau, M. (1985) 'The Practices of Space', in *On Signs* M. Blonsky (ed.), *On Signs*. Oxford: Blackwell. pp. 122–45.
Cullen, I. (1978) 'The Treatment of Time in the Explanation of Spatial Behaviour', (in T. Carlstein, D. Parkes and N. Thrift, (eds), *Timing Space and Spacing Time*. Vol. 2: *Human Activity and Time Geography*. London: Arnold. pp. 27–39.
Currie, G. (1999) 'Can There Be a Literary Philosophy of Time?' in J. Butterfield (ed.), *The Arguments of Time*. Oxford: Oxford University Press. pp. 43–63.
Deleuze, G. (1986) *Cinema 1: The Movement-Image*. Minneapolis: University of Minnesota Press.
Deleuze, G. (1989) *Cinema 2: The Time-Image*. Minneapolis: University of Minnesota Press.

Deleuze, G. (1991) *Bergsonism*. New York: Zone Books.

Eco, U. (1989) *The Middle Ages of James Joyce: The Aesthetics of Chaosmos*. London: Hutchinson Radius.

Eisenman, P. (1997) 'Separate Tricks', in J. Kipnis and T. Leeser (eds), *Chora L Works: Jacques Derrida and Peter Eisenman*. New York: Monacelli Press. pp. 132–6.

Elden, S. (2001) *Mapping the Present: Heidegger, Foucault and the Project of a Spatial History*. London: Continuum.

Felski, R. (2000) 'The Invention of Everyday Life', *New Formations* 39: 13–32.

Fenn, R. (2001). *Time Exposure: The Personal Experience of Time in Secular Societes*. Oxford: Oxford University Press.

Flaherty, M. (1999) *The Watched Pot: How We Experience Time*. New York: New York University Press.

Foucault, M. (1986) 'Of Other Spaces', *Diacritics* 16 (1): 22–7.

Goff Le, J. (1980) *Time and Culture in the Middle Ages*. Chicago, IL: University of Chicago Press.

Griffiths, J. (1999) *Pip Pip: A Sideways Look at Time*. London: Flamingo.

Gross, D. (1985) 'Bergson, Proust and the Revaluation of Memory', *International Philosophical Quarterly* 25: 369–80.

Grosz, E. (1995) *Space, Time and Perversion: Essays on the Politics of Bodies*. London: Routledge.

Guattari, F. (1992) *Chaosmosis: A New Ethico-Aesthetic Paradigm*. Bloomington, IN: Indiana University Press.

Hägerstrand, T. (1982) 'Diorama, Path, Project', *Tijdschrift voor Economische en Sociale Geografie* 73 (6): 323–39.

Hall, E.T. (1983) *The Dance of Life: The Other Dimension of Time*. New York: Doubleday.

Harvey, D. (1989) *The Condition of Postmodernity*. Oxford: Blackwell.

Holquist, M. (1990) *Dialogism: Bakhtin and His World*. London: Routledge.

Jameson, F. (1998) *The Cultural Turn: Selected Writings on the Postmodern, 1983–1998*. London: Verso.

Jameson, F. (1999) Time and the Concept of Modernity, in C. Davidson (ed.), *Anytime*. Cambridge, MA: MIT Press. pp. 208–17.

Jameson, F. (2003) 'The End of Temporality', *Critical Inquiry* 29 (Summer): 695–718.

Johnson, J. (2000) 'Literary Geography: Joyce, Woolf and the City', *City: Analysis of Urban Trends, Culture, Theory, Policy, Action* 4 (2): 199–214.

Kneale, J. and Holloway, J. (2000) 'Mikhail Bakhtin', in M. Crang and N. Thrift (eds), *Thinking Space*. London: Routledge. pp. 71–88.

de Landa, M. (1998) 'Extensive Borderlines and Intensive Borderlines', in L. Woods and E. Rehfeld (eds), *Borderline*. New York: Springer. pp. 18–24.

de Landa, M. (1999) 'Deleuze, Diagrams and the Open-Ended Becoming of the World', in E. Grosz (ed.), *Becomings: Explorations in Time, Memory and Futures*. Ithaca, NY: Cornell University Press. pp. 29–41.

Lefebvre, H. (1991) *The Production of Space*. Oxford: Blackwell.

Linklater, A. (2002) *Measuring America: How the United States Was Shaped by the Greatest Land Sale in History*. London: Harper Collins.

Lloyd, G. (1993) *Being in Time: Selves and Narrators in Philosophy and Literature*. London: Routledge.

Maffesoli, M. (1998) 'Presentism – or the Value of the Cycle', in S. Lash, A. Quick and R. Roberts (eds), *Time and Value*. London: Sage. pp. 103–12.

Massey, D. (1992) 'Politics and Space/Time', *New Left Review* (196): 65–84.

Massey, D. (1998) 'Philosophy and the Politics of Spatiality: power-geometries and the politics of space-time', Hettner Lecture at the University of Heidelberg.

Maynes, M. (1989) 'Gender and Narrative Form in French and German Working-Class Autobiographies', in Personal Narratives Group (eds), *Interpreting Women's Lives: Feminist Theory and Personal Narratives*. Bloomington, IN: Indiana University Press. pp. 103–17.

Maynes, M. (1992) 'Autobiography and Class Formation in Nineteenth Century Europe: Methodological Considerations', *Social Science History* 16 (3): 517–37.

Nowotny, H. (1994) *Time: The Modern and Postmodern Experience*. London: Polity Press.

Perkins, M. (1998) 'Timeless Cultures – the "Dreamtime" as Colonial Discourse', *Time & Society* 7 (2): 335–51.

Poulet, G. (1977) *Proustian Spaces*. Baltimore, MD: Johns Hopkins University Press.

Rämö, H. (1999) 'An Aristotelian Human Time–Space Manifold: From *Chronochora* to *Kairotopos*', *Time & Society* 8 (2): 309–28.

Rickard, J. (1999) *Joyce's Book of Memory: The Mnemotechnic of Ulysses*. Durham, NC: Duke University Press.

Ricoeur, P. (1988) *Time and Narrative*, Vol. 3. Chicago: University of Chicago Press.

Schleiffer, R. (2000) *Modernism and Time: The Logic of Abundance in Literature, Science and Culture, 1880–1920*. Cambridge: Cambridge University Press.

Soja, E.W. (1989) *Postmodern Geographies: The Reassertion of Space in Critical Social Theory*. London: Verso.

Spivak, G.C. (1991) 'Time and Timing: Law and History', in J. Bender and D. Wellberg (eds), *Chronotypes: The Construction of Time*. Stanford, CA: Stanford University Press. pp. 99–117.

Tschumi, B. (1999) Diasync in C. Davidson (ed.), *Anytime*. Cambridge, MA: MIT Press. pp. 168–75.

Wigley, M. (1993) *The Architecture of Deconstruction: Derrida's Haunt*. Cambridge, MA: MIT Press.

Index

Indexed by Caroline Eley